10 Chemistry of Plant Protection

Herbicides Inhibiting Branched-Chain Amino Acid Biosynthesis - Recent Developments

Volume Editor: J. Stetter

With contributions by
H. M. Brown, J. C. Cotterman, M. W. Drewes,
G. J. De Boer, S. K. Gee, B. C. Gerwick, J. V. Hay,
W. A. Kleschick, D. W. Ladner, F. Lieb,
U. Philipp, J. V. Schloss, P. R. Schmitzer

With 112 Figures and 18 Tables

Springer-Verlag
Berlin Heidelberg NewYork
London Paris Tokyo
HongKong Barcelona Budapest

Volume Editor:

Prof. Dr. *J. Stetter*
ZF - FWI, Bayer AG, Q 18
51368 Leverkusen, Bayerwerk/FRG

This series continues the handbook
"Chemie der Pflanzenschutz- und Schädlingsbekämpfungsmittel"
edited by R. Wegler

ISBN-13: 978-3-642-79109-3 e-ISBN-13: 978-3-642-79107-9
DOI: 10.1007/978-3-642-79107-9

Library of Congress Cataloging-in-Publication Data
Herbicides inhibiting branched-chain amino acid biosynthesis : recent developments / volume editor, J. Stetter ; with contributions by H.M. Brown ... [et al.]. p. cm. -- (Chemistry of plant protection ; 10) Included index.
ISBN-13: 978-3-642-79109-3

1. Herbicides. 2. Branched chain amino acids--Synthesis. 3. Branched chain amino acids--Synthesis--Inhibitors. J. Stetter, J. (Jörg) II. Brown, H. M. III. Series.
SB951.4.H465 1994 668'.654--dc20 94-34261 CIP

Softcover reprint of the hardcover 1st edition 1994

Typesetting: Macmillan India Ltd., Bangalore-25
SPIN: 10123402 52/3020 - 5 4 3 2 1 0 - Printed on acid-free paper

Managing Editors

Editorial Board

Editorial

The managing editors and I are proud that ten volumes of the book series "Chemistry of Plant Protection" have been published before a decade has passed since the start of this project in 1986.

The objectives of this enterprise are still the same: To supply readers with reviews about scientific fields of modern interest; the reviews are directed to new developments in research of all questions regarding chemical plant protection in its broadest sense. This includes synthesis and properties of new biochemically active substance classes, their trace analysis in environmental specimens, the environmental influence and behaviour of pesticides as well as compounds attributed with properties generating and controlling biological processes which may protect nutritional plants against diseases or other influences detrimental to growth.

In order to reach a wide readership all papers are written in English. The publisher and the board endeavour to engage highly specialized scientists for the respective articles and it is self-evident that only high qualified contributions will be accepted.

The contributions are written to a fairly advanced level and should be of interest to graduate students, research fellows and practising scientists. I am sure that the subject matter treated is of high interest to scientists in industry, universities, as well as in governmental and regulatory bodies. In addition, natural products chemists, biologists, ecologists, and organic chemists will benefit from several of the chapters within this book series.

It is with great satisfaction that I thank all authors and editors involved for the time they have devoted to this effort.

Berlin, October 1994

W. Ebing
Editor-in-Chief

Table of Contents

Introduction

J. Stetter

Despite some interesting solutions that have emerged in the field of biological crop protection, chemical crop protection will still play a dominant role well into the next century. Most experts agree on this assessment. Chemical crop protection is based largely on synthetic chemistry; natural products have been used in practice only in rare cases (pyrethrum, neem extract, nicotine, rotenone, etc.).

Synthetic chemistry has far from shot its bolt. It is still producing a constant stream of new lead structures which are certainly worth following up, as the current patent situation clearly shows.

One of the aims of the "Chemistry of Plant Protection" series is to introduce major, pioneering active ingredient innovations to a wider audience. Two classes of active ingredients which have each set new standards in the elimination of plant diseases and harmful insects have already been covered: fungicidal ergosterol biosynthesis inhibitors (volume 1) and the insecticidal pyrethroids (volumes 4 and 5). The idea for a volume on "Herbicides inhibiting branched chain amino acid biosynthesis" was born early in 1990. In a survey of colleagues in crop protection research, this topic was mentioned most frequently as a possible idea.

This is hardly surprising, since products from the sulphonylurea and imidazolinone classes have brought about almost revolutionary progress in practice. The products introduced to the market so far are based on no less than 17 different ALS (acetolactate synthase) active ingredients. They display selectivities in and applicability to all major crops.

In 1990 a very good review on ALS inhibitors was published [1]. Other contributions on special topics followed soon thereafter [2–4]. So the original plan of producing a comprehensive review on ALS inhibitors was modified in favour of a progress report on recent developments in this rapidly changing field.

The herbicidal sulphonylureas could really have been discovered ten years earlier. A 1966 patent described the herbicidal properties of sulphonylureas with an aminotriazine substructure [5]. As derivatives of photosynthesis inhibitors such as propazine, however, their activity was only about the same as the compound from which they were derived. A slight shift to differently substituted aminotriazines would have been enough to indicate the potency of the ALS sulphonylureas even then. It is ironic that a pioneering discovery was missed by a cat's whisker, but it just goes to show how much success in active ingredient research often depends on pure chance.

It was not until the middle of the 1970s that G. Levitt at DuPont turned his attention to sulphonylureas, at first more out of a purely chemical interest. It was his serendipity, that happily led him to follow up a moderately effective lead structure [6].

It was an astounding coincidence that the imidazolinones were discovered almost simultaneously at American Cyanamid in the late 1970s [7]. Although this second class is structurally very different it acts on the same target enzyme.

The 1980s saw the start of intensive research all over the world into ALS inhibitors. Practically all the major crop protection companies were involved in this work, as the patent publications disclosed to date show. The structures which have so far been acknowledged as effective range from close analogues of the sulphonylureas and imidazolinones to triazolopyrimidines and salicylic acid derivatives whose structural similarity to each other is almost impossible for the non-specialist to discern.

There is now an increasing number of patent applications for novel herbicidal structures which, on closer examination by an expert, would appear to be ALS inhibitors.

The present volume attempts to give an overview of the field as up-to-date as possible. In the chapters on the biochemistry of ALS, on the sulphonylureas, imidazolinones and triazolopyrimidines, distinguished experts from the inventing companies report on recent developments and findings. The state-of-the-art contributions on salicylic acid derivatives and the broad spectrum of other ALS structures were written by scientists who themselves are involved in these fields and have a profound knowledge of the material.

This volume can do no more than represent the situation at a given point in time as the field is still very much in flux. It is almost impossible to say where the journey of scientific discovery will take us next and what other innovative possibilities ALS research will offer us.

Perhaps something will happen which is in fact not so uncommon in active ingredient research: a switch of indication. The screening procedures used in pharmaceutical and crop protection research are nowadays so diversified and of such high capacity that the wide-ranging field of ALS chemistry with its potential of molecular diversity could well produce positive responses for completely different targets. We will just have to wait and see, because only the future will tell.

References

1. Pestic. Sci. (1990) 29: 241
2. Baker DR, Fenyes JG, Moberg WK (eds) (1991) Synthesis and chemistry of agrochemicals II. American Chemical Society, Washington DC, ACS Symposium Series No. 443, 16
3. Baker DR, Fenyes JG, Steffens JJ (eds) (1992) Synthesis and chemistry of agrochemicals III. American Chemical Society, Washington DC, ACS Symposium Series No. 504, p 10
4. Shaner DL, O'Connor SL (eds) (1991) The imidazolinone herbicides, CRC Press
5. Koog HJ (1966) Netherlands Patent 121788
6. Sauers RF, Levitt G (1984) In: Magee PS, Kohn GK, Menn JJ (eds) Pesticide synthesis through rational approaches. American Chemical Society, Washington DC, ACS Symposium Series No. 255, p 21
7. Los M (1984) In: Magee PS, Kohn GK, Menn JJ (eds) Pesticide synthesis through rational approaches, American Chemical Society, Washington DC, ACS Symposium Series No. 255, p 29

Recent Advances in Understanding the Mechanism and Inhibition of Acetolactate Synthase

John V. Schloss

Department of Medicinal Chemistry, 4070 Malott Hall,
The University of Kansas, Lawrence, KS 66045-2506, USA

A review of our current understanding of the structure of acetolactate synthase, the molecular mechanism of its inhibition by herbicides, and how this inhibition is translated into phytotoxicity in vivo, is presented.

Determination of the crystal structures of the three thiamine pyrophosphate-dependent enzymes, transketolase, pyruvate oxidase, and pyruvate decarboxylase, have provided the first insight into the likely three-dimensional structure of acetolactate synthase.

Discovery of the oxygenase activity of the bacterial acetolactate synthase has provided a rationale for irreversible losses of enzyme activity encountered with various herbicides.

The reasons for potent bacterial stasis and herbicidal effects as a consequence of the inhibition of acetolactate synthase have only become less clear. Buildup of ketobutyrate, once thought to play a major role in the physiological consequences of this inhibition, does not, per se, give rise to these effects in either plants or bacteria. Starvation of plants for branched chain amino acids also does not seem to correlate well with the phenotypic response, leaving the physiology associated with inhibition of acetolactate synthase more of a mystery than ever.

Chemistry of Plant Protection, Vol. 10

1 Structure of ALS

Acetolactate synthase (ALS, EC 4.1.3.18, also commonly referred to as acetohydroxy acid synthase) is the first common enzyme of branched chain amino acid biosynthesis. The two physiological reactions catalyzed by ALS are the condensation of two molecules of pyruvate to form *S*-acetolactate (the "aceto" moiety from the first pyruvate and the "lactate" moiety from the second) and carbon dioxide (from the first pyruvate), or the condensation of pyruvate and α-ketobutyrate to form α-aceto-α-hydroxybutyrate (the "aceto" moiety from the pyruvate and the "hydroxybutyrate" moiety from the ketobutyrate) and carbon dioxide (from the pyruvate) [1, 2]. These two reactions are specific to the biosynthesis of valine (or leucine) and isoleucine, respectively. The enzyme devoted to the biosynthesis of branched chain amino acids is a flavoprotein (bacteria, yeast, and higher plants), and has an absolute requirement for flavin adenine dinucleotide, thiamine pyrophosphate, and a di- or trivalent metal for activity [3]. The enzyme isolated from bacteria is composed of two types of subunits and exists in an $\alpha_2\beta_2$ oligomeric state (140,000 kDa); pyruvate induces a higher oligomer to form, $\alpha_4\beta_4$ ([4] Van Dyk and Schloss, unpublished observation). The enzyme isolated from yeast or plants consists of a single type of polypeptide (ca. 60,000 kDa) and exists as an octomer or a tetramer (ca. 240,000 kDa, in the presence of pyruvate or when stripped of its flavin adenine dinucleotide cofactor) [5, 6].

Acetolactate synthase isozymes from bacteria, yeast, and higher plants share sequence homology with several mechanistically related enzymes: pyruvate decarboxylase (EC 4.1.1.1), the quinone-dependent pyruvate oxidase (EC 1.2.2.2) and the O_2-linked, acetyl phosphate-forming pyruvate oxidase (EC 1.2.3.3) [3, 7, 8, 9]. Although a crystal structure for acetolactate synthase from any source has yet to be determined, structures of the O_2-dependent pyruvate oxidase [8] and pyruvate decarboxylase [10] in addition to the related thiamine pyrophosphate-dependent enzyme transketolase [11], have recently been reported. Like acetolactate synthase, the O_2-dependent pyruvate oxidase requires both a flavin adenine dinucleotide and thiamine pyrophosphate cofactor, and assembles into a tetrameric oligomeric structure. However, the O_2-dependent pyruvate oxidase differs from acetolactate synthase and the quinone-dependent pyruvate oxidase in that it transfers electrons efficiently from pyruvate to molecular oxygen. Presumably, the O_2-dependent oxidase does not bind quinones (e.g. Q_0), as do acetolactate synthase and the other pyruvate oxidase. Thiamine pyrophosphate is bound in an extended conformation in all three crystal structures with the metal coordinating the pyrophosphate moiety of the cofactor and serving to anchor it to the protein through an interaction with a specific region of protein structure (X_5GlyAspGlyY_{25-28}Asn). Similarly, in pyruvate oxidase the flavin adenine dinucleotide is bound in an extended conformation, with the isoalloxazine moiety (xylene portion) oriented closest to C-2 of the thiazole ring of thiamine pyrophosphate. Although a hydride is thought to be transferred between the hydroxyl of the hydroxyethyl-thiamine pyrophosphate reaction intermediate and N-5 of the flavin's isoalloxazine ring in pyruvate oxidase, the distance between these two points (9.5 Å) would seem to be too great to facilitate a hydride transfer mechanism [8]. Possibly, pyruvate oxidase and acetolactate synthase share a similar mechanism for interaction with molecular oxygen that does not involve hydride transfer to the flavin, as outlined in the next section. The thiamine pyrophosphate cofactor is bound in all three enzyme crystal structures with its C-4′ amino group adjacent to C-2 of the thiazole ring [8, 10, 11]. This conformation would be consistent with a role for the C-4′ amino group in facilitating ionization of the cofactor's thiazole [12, 13].

Unfortunately, the question of greatest interest with respect to these crystal structures and the herbicides that inhibit acetolactate synthase remains to be answered: what is the nature of the herbicide binding site? Although the quinone-dependent pyruvate oxidase and acetolactate synthase both bind quinones (Q_0 and Q_1), the former enzyme is not inhibited by the commercial herbicides that are potent inhibitors of acetolactate synthase (nor has its crystal structure been solved) [14]. Even among the three isozymes of acetolactate synthase from enteric bacteria, there is a wide variance in their sensitivity to different herbicides [15]. Those changes in the quinone-binding domain that have occurred between pyruvate oxidase and acetolactate synthase are of upmost importance in any effort to improve herbicidal potency by reference to crystal structure. For this reason, it is likely that the structure of acetolactate synthase from any source other than a higher plant will be of limited value in guiding future synthetic efforts devoted to design of new herbicides. Despite this current limitation, the initial speculation that the herbicide site of acetolactate synthase is derived, evolutionarily speaking, from the quinone site of pyruvate oxidase, remains as the most coherent explanation as to the nature of this site [15]. Although this proposal has been challenged based on the differential effect that some mutations have on one class of herbicides relative to another [16, 17], results to date are consistent with a binding domain common to all classes of herbicides and smaller subdomains that are unique to chemically distinct subclasses of structures [18, 19]. Although herbicides can also compete with binding of branched chain amino acids for a "regulatory site" on acetolactate synthase [19, 20], binding of branched chain amino acids is nearly competitive with pyruvate (and hyperbolic, [21]), as are, to a large extent, many of the herbicides [15], and the isozyme of acetolactate synthase in enteric bacteria most sensitive to inhibition by commercial herbicides, isozyme II, is completely insensitive to inhibition by branched chain amino acids.

2 Irreversible Inhibition and In Vivo Potency

Inhibition of acetolactate synthase by herbicides is kinetically very complex [15]. In addition to a reversible, time-dependent inhibition of the enzyme, that depends on enzymic turnover for potent interaction between enzyme and herbicides, there is also an irreversible loss of enzymic activity at a somewhat slower rate [22]. The rate of irreversible inactivation of the enzyme must differ for different classes of herbicides, at least in vivo, as treatment of plants with imidazolinone herbicides leads to irreversible loss of extractable acetolactate synthase and sulfonylurea herbicides can prevent the irreversible loss of activity [23, 24]. Irreversible loss of enzymic activity in vivo is not a consequence of proteolytic degradation of the enzyme or covalent attachment of the herbicide [21, 25, 26].

The nature of this irreversible inactivation of acetolactate synthase by herbicides is likely to be associated with the recently discovered oxygen consuming side reaction of this enzyme [27]. The second pyruvate or ketobutyrate competes with molecular oxygen for the ene-amine form of hydroxyethyl-thiamine pyrophosphate (functionally a carbanion). A likely mechanism for the oxygenase reaction of acetolactate synthase is illustrated in Fig. 1. This differs from the presumed mechanism of the acetyl phosphate-forming pyruvate oxidase. By contrast to acetolactate synthase, oxidation of the hydroxyethyl-

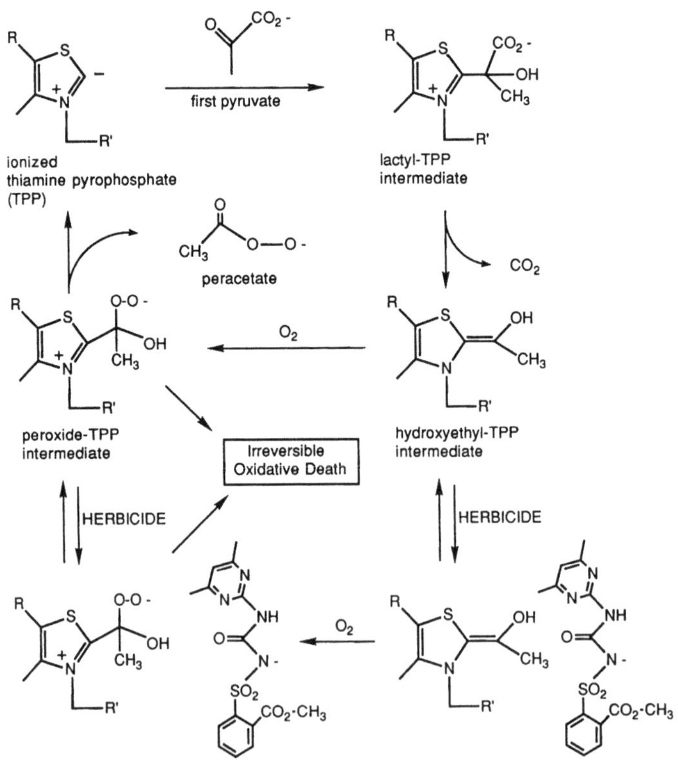

Fig. 1. The mechanism of the oxygenase reaction of acetolactate synthase and the role of herbicide (sulfometuron methyl) binding in potentiating oxidative inactivation of the enzyme

thiamine pyrophosphate reaction intermediate would give reduced flavin adenine dinucleotide, followed by oxidation of the flavin by molecular oxygen. Evidence of different mechanisms for the oxidase and oxygenase reactions are the effect that reconstitution of these enzymes with 5-deaza-flavin adenine dinucleotide has on the rate of these reactions. The 5-deaza flavin-reconstituted pyruvate oxidase is virtually inactive [28], by contrast to acetolactate synthase, that has an identical level of oxygenase activity to the flavin-activated enzyme [3]. These results strongly suggest that the flavin plays no role in the oxygenase reaction of acetolactate synthase (as illustrated in Fig. 1), but is likely to interact directly with oxygen in the much faster oxygen-consuming reaction of pyruvate oxidase.

For acetolactate synthase, addition of the first pyruvate occurs at low concentrations ($< 10\,\mu M$) and is highly committed to decarboxylation. Ketobutyrate or higher concentrations of pyruvate inhibit the oxygenase reaction of acetolactate synthase, in contrast to acetolactate, that does not inhibit the oxygenase reaction and is required at higher concentrations to saturate this reaction [29]. The form of the enzyme that reacts with molecular oxygen is the same that binds herbicides most tightly, the complex with the ene-amine form of hydroxyethyl-thiamine pyrophosphate. Herbicides will bind tightly to this form of the enzyme generated from pyruvate or acetolactate, completely inhibiting the rate of oxygen consumption [30]. The effect of herbicides on the oxygenase rate of acetolactate synthase is in contrast to the effect that high concentrations of pyruvate or ketobutyrate have, incomplete inhibition [29]. This quantitative difference in the effect of α-oxo acids and herbicides can be explained by the inability of α-oxo acids to prevent release of the product of the oxygenase reaction, peracetate [27]. By contrast, herbicides inhibit the synthase and oxygenase reactions of acetolactate synthase by a similar mechanism, they prevent the release of products by binding tightly to the intermediate form of the enzyme (Fig. 1), even after addition of the second α-oxo acid or oxygen. Here, there is a quantitative difference between the sulfonylurea and imidazolinone herbicides that is consistent with their relative tendencies to cause inactivation of the enzyme, the imidazolinones bind as tightly or more tightly after addition of the second substrate to the enzyme, while the sulfonylureas bind substantially less tightly after addition of the second substrate (high concentrations of pyruvate, ketobutyrate, or addition of oxygen) [15]. The incomplete inhibition of the oxygenase reaction by high concentrations of α-oxo acid indicates that oxygen has more than one route of access to the carbanionic reaction intermediate of acetolactate synthase [29]. If the carbanionic intermediate trapped in the herbicide complex were similarly accessible to oxygen, then herbicides would hold the enzyme in an oxygen-sensitive state and retard release of the peroxide product. Retention of the peroxide on the enzyme by complexation with herbicide would increase the likelihood of oxidative inactivation of the enzyme. An oxidative death for acetolactate synthase would be consistent with the observed release of herbicide without recovery of enzymic activity [22, 26] and retention of a native molecular weight [25] by the inactive enzyme.

In the presence of oxygen and pyruvate (or acetolactate), acetolactate synthase will generate the highly reactive peracetate [27]. Loss of enzymic activity under these conditions, however, is probably not due to a buildup of exogenous oxidant, as dithiothreitol cannot protect the enzyme against inactivation [31]. It would appear that inactivation of the enzyme under these conditions is due to oxidative damage to the protein by the enzyme-bound peroxide intermediate, peracetate, or other reactive oxygen species prior to their release by acetolactate synthase. In the absence of pyruvate, acetolactate synthase also loses activity at a somewhat slower rate after activation by a divalent metal and thiamine pyrophosphate [31]. Although this inactivation in the absence of pyruvate could be due to reaction of oxygen with the carbanion formed by ionization at C-2 of the thiazole of thiamine pyrophosphate, it has not been established that inactivation under these conditions requires oxygen. Under non-turnover conditions (no α-oxo acid or acetolactate) the divalent metal required for binding of the thiamine pyrophosphate cofactor [14] can inactivate acetolactate synthase. After incubation of stoichiometric Co^{2+}, thiamine pyrophosphate, and flavin adenine dinucleotide, with acetolactate synthase isozyme II from *Salmonella typhimurium* (ca. 1 mM protomer) for 12 h at 25 ° C in the dark, precipitation of the enzyme with irreversible loss of enzymic activity occurs (Schloss, unpublished observation). Even in the absence of thiamine pyrophosphate, acetolactate synthase is sensitive to inactivation by various metals. Clearly, attributing the loss of enzymic activity to a particular cause under conditions of aerobic turnover must be done with caution.

Acetolactate synthase is one of several carbanion-forming enzymes that have oxygen consuming side reactions. Other enzymes that have oxygenase activities include ribulosebisphosphate carboxylase, glutamate decarboxylase from *Escherichia coli*, fructose 1,6-bisphosphate aldolase from *Staphylococcus aureus*, and the mechanistically related enzyme pyruvate decarboxylase [27]. The oxygenase activity of ribulosebisphosphate carboxylase is responsible for oxygen inhibition of photosynthesis, or the Warburg effect [32]. Unlike acetolactate synthase, the oxygenase activity of ribulosebisphosphate carboxylase does not normally give rise to a reactive oxygen species, but involves efficient cleavage of the oxygen–oxygen bond of the peroxide intermediate [33]. Also in contrast to the likely role of metal in the oxygenase reactions of acetolactate synthase or pyruvate decarboxylase, the activating metal of ribulosebisphosphate carboxylase seems to play a very intimate role in determining the ratio between the carboxylase and oxygenase activities [34]. The ribulosebisphosphate carboxylase from *Rhodospirillum rubrum* is an exclusive oxygenase when activated by Co^{2+}. The relative levels of synthase and oxygenase activities for acetolactate synthase vary by less than six-fold with an extremely diverse set of activating metals (Mn^{2+}, Mg^{2+}, Ca^{2+}, Co^{2+}, Zn^{2+}, Ni^{2+}, Cu^{2+}, Ba^{2+}, Cd^{2+}, and Al^{3+}), consistent with the more subtle role for metal of thiamine pyrophosphate binding [29].

When activated with Mn^{2+}, the oxygenase activity of ribulosebisphosphate carboxylase has an associated chemiluminescence that has been attributed to the formation of singlet oxygen [35]. Acetolactate synthase isozyme II from *Salmonella typhimurium* has comparable levels of chemiluminescence to that seen with ribulosebisphosphate carboxylase (Schloss, unpublished observation). Chemiluminescence has also been observed for acetolactate synthase isolated from corn, and is dependent on pyruvate, inhibitable by chlorsulfuron, and quenched by the singlet oxygen scavenger sodium azide [36]. The chemiluminescence observed for acetolactate synthase isolated from a higher plant [36] provides evidence that it has an oxygenase activity similar to that seen for isozymes I and II of this enzyme from enteric bacteria [27, 29]. If the chemiluminescence observed for these enzymes is due to the formation of singlet oxygen, there are at least two possible explanations for how it may be formed. As illustrated in Fig. 2, reaction of triplet oxygen with the ene-amine form of the hydroxyethyl-thiamine pyrophosphate reaction intermediate probably occurs in two steps. Transfer of an electron from the carbanion to oxygen would give a carbon radical and superoxide anion. Reaction of superoxide with the radical intermediate would give a peroxide anion. This peroxide intermediate could eliminate the ionized thiazole of thiamine pyrophosphate to give the peracetate product, or could regenerate the carbanionic intermediate by a hysteretic reversal of the reaction to give singlet oxygen (no longer a "spin forbidden" process). Under assay conditions, two molecules of peracetate rapidly dismutate to form two molecules of acetate and oxygen [27]. If this process were to give

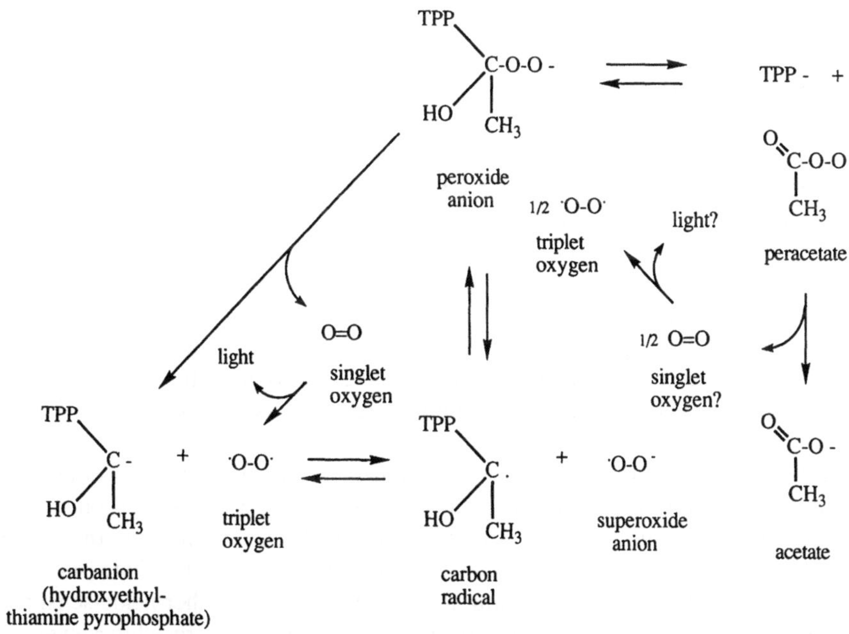

Fig. 2. Reaction of oxygen with the carbanionic reaction intermediate of acetolactate synthase and possible products and intermediates of this reaction

singlet oxygen as the product, then this could account for the chemiluminescence observed. Evidence for peracetate as an intermediate in the formation of acetate was provided by the incorporation of one atom of ^{18}O-label from $^{18}O_2$ into the carboxyl of acetate and the lack of any incorporation from $H_2{}^{18}O$ [27]. Peracetate could also be directly observed by its oxidation of *o*-dianisidine catalyzed by horse radish peroxidase, despite it being a rather poor substrate for this enzyme (Abell and Schloss, unpublished observation). Release of any of the intermediates of the oxygenase reaction, as outlined in Fig. 2, could give superoxide, carbon radicals, alkyl peroxides, or singlet oxygen as products. Superoxide, if formed, will give hydrogen peroxide [37] and singlet oxygen in a water dependent reaction [38]. The origin of the chemiluminescence associated with acetolactate synthase is not clear. If it is due to singlet oxygen, then the visible emissions (580, 634, and 703 nm) are due to dimol decay, or the simultaneous conversion of two molecules of singlet oxygen to triplet oxygen (a relatively rare event, i.e. very little light is produced). The primary emissions for conversion of singlet to triplet oxygen that occur in the infrared (1070, 1580, and 1268 nm, with the latter emission of greatest intensity [39]) have yet to be characterized. If these emissions prove to be due to singlet oxygen, then it still would not be clear whether singlet oxygen were formed by hysteretic reversal of the oxygenase reaction, reactions involving peracetate, or indirectly via superoxide. Clearly, the oxygenase reactions of acetolactae synthase and the other carbanion-forming enzymes are potential sources of a variety of oxidants.

3 Physiological Consequences of ALS Inhibition

Although inhibition of acetolactate synthase in bacteria causes stasis, in plants inhibition of this enzyme is lethal. The reasons for this lethality in plants and the potent stasis in bacteria have been topics of a great deal of conjecture. In bacteria, several lines of evidence support the notion that buildup of one of the substrates for acetolactate synthase, α-ketobutyrate, contributes substantially to the potency of stasis [40–46]. In mutants that have lost feedback inhibition of threonine deaminase (the enzyme that makes ketobutyrate) by isoleucine, inhibition of growth by herbicides cannot be reversed as readily by isoleucine as in wild type organisms [42,43]. Mutants that have lost or reduced the levels of other enzymes that utilize ketobutyrate (pyruvate oxidase) or the products of ketobutyrate catabolism, such as propionyl-CoA (acetate kinase and phosphotransacetylase), exhibit hypersensitivity to herbicides, as do mutants that have reduced capacity to synthesize coenzyme A or thiamine pyrophosphate [44–46].

However, in a bacterial strain that expresses only a herbicide-sensitive isozyme of acetolactate synthase (isozyme II), elevation of intracellular levels of α-ketobutyrate occurs only at concentrations well above the concentrations of herbicide required to inhibit growth [47]. Further, at concentrations of herbicide that will inhibit the rate of growth of this strain of bacteria, but not wild type *Salmonella typhimurium* (due to the presence of the less sensitive isozyme I), the levels of α-ketobutyrate are elevated in both organisms to the same extent, despite the fact that growth inhibition is observed with only the former strain. A strain of *S. typhimurium* that contains only isozyme I of acetolactate synthase grows slowly, presumably due to its inefficiency in synthesizing isoleucine. This strain has elevated levels of α-ketobutyrate relative to wild type, but addition of exogenous α-ketobutyrate stimulates the rate of growth despite elevating intracellular α-ketobutyrate levels even further. *Chlorella emersonii* is sensitive to inhibition by sulfometuron methyl. Inhibition of this organism is associated with a transient elevation of α-ketobutyrate, but growth inhibition and the inhibition of protein and chlorophyll synthesis persist after the intracellular levels of α-ketobuyrate have returned to normal.

Although α-ketobutyrate and α-aminobutyrate are elevated in corn seedlings treated with the herbicide imazaquin, concurrent treatment of plants with isoleucine will reduce these levels to near normal without affecting the inhibitory effect of the herbicide [48]. Further the levels of valine and leucine in plants treated with a combination of imazaquin and isoleucine are about half of what is obtained by treatment of plants with imazaquin alone. Treatment of plants with a combination of imazaquin and valine led to a dramatic reduction in the level of isoleucine (valine and leucine were elevated), in contrast to plants treated with imazaquin alone where the level of isoleucine was elevated. While these results provide convincing evidence that growth inhibition in plants is not a consequence of α-ketobutyrate buildup by itself, they do not provide a convincing argument that growth inhibition is due to starvation for branched chain amino acids.

Unfortunately, we are presently left with an open question rather than a definitive answer to the reasons for stasis in bacteria and death in plants induced by inhibitors of acetolactate synthase. It seems clear that elevation of α-ketobutyrate or reduction of the levels of branched chain amino acids alone will not give these effects. There are quantitative differences in the degree of inhibition required to give toxicity for the three common enzymes of branched chain amino acid biosynthesis [18]. Whether the underlying reasons for these differences are simply due to the relative abundance of these activities compared to the level required for normal function, secondary effects due to perturbations in pathway flux at that point, or a propitious combination of these effects, remain to be fully elucidated.

4 References

1. Crout DHG (1990) The chemistry of branched chain amino acids biosynthesis: stereochemical and mechanistic aspects, In: Barak Z, Chipman DM, Schloss JV (eds) Biosynthesis of Branched Chain Amino Acids. VCH, Weinheim, pp 199–242

2. Crout DHG, Lee ER, Rathbone DL (1990) Absolute configuration of the product of the acetolactate synthase reaction by a novel method of analysis using acetolactate decarboxylase, J Chem Soc, Perkin Trans 1, 1367–1369
3. Schloss JV (1992) Acetolactate synthase, In: Müller F (ed) Chemistry and Biochemistry of Flavoenzymes, Vol. III, pp 531–542
4. Grimminger H, Umbarger HE (1979) Acetohydroxy acid synthase I of *Escherichia coli*, purification and properties, J Bacteriol 137, 846–853
5. Durner J, Böger P (1990) Oligomeric forms of plant acetolactate synthase depend on flavin adenine dinucleotide, Plant Physiol 93, 1027–1031
6. Singh BK, Newhouse KE, Stidham MA, Shaner DL (1990) Imidazolinones and acetohydroxyacid synthase from plants. In: Barak Z, Chipman DM, Schloss JV (eds) Biosynthesis of Branched Chain Amino Acids, pp 357–371
7. Koland JG, Miller MJ, Gennis RB (1984) Reconstitution of the membrane bound ubiquinone dependent pyruvate oxidase respiratory chain of *Escherichia coli* with cytochrome D terminal oxidase, Biochemistry 23, 445–453
8. Muller YA, Schulz GE (1993) Structure of the thiamine- and flavin-dependent enzyme pyruvate oxidase, Science 259, 965–967
9. Green JBA (1989) Pyruvate decarboxylase is like acetolactate synthase (ILV2) and not like pyruvate dehydrogenase E1 subunit, FEBS Lett 246, 1–5
10. Dyda F, Furey W, Swaminathan S, Sax M, Farrenkopf B, Jordan F (1993) Catalytic centers in the thiamine diphosphate dependent enzyme pyruvate decarboxylase at 2.4 Å resolution, Biochemistry 32, 6165–6170
11. Lindqvist Y, Schneider G, Ermler U, Sundstroem M (1992) Three-dimensional structure of transketolase, a thiamine diphosphate dependent enzyme, at 2.5 Å resolution, EMBO J 11, 2373–2379
12. Kluger R (1987) Thiamine diphosphate: a mechanistic update on enzymic and nonenzymic catalysis of decarboxylation, Chem Rev 87, 863–876
13. Suchy J, Mieyal JJ, Bantle G, Sable HZ (1972) Coenzyme interactions. VI Properties of the pyrimidine moiety of thiamine, J Biol Chem 247, 5905–5912
14. Schloss JV, Aulabaugh A (1990) Acetolactate synthase and ketol-acid reductoisomerase: a search for reason and a reason for search. In: Barak Z, Chipman DM, Schloss JV (eds) Biosynthesis of Branched Chain Amino Acids. VCH, Weinheim, pp 329–356
15. Schloss JV, Ciskanik LM, Van Dyk DE (1988) Origin of the herbicide binding site of acetolactate synthase, Nature (London) 331, 360–362
16. Sathasivan K, Haughn GW, Murai N (1991) Molecular basis of imidazolinone herbicide resistance in *Arabidopsis thaliana* var *Columbia*, Plant Physiol 97, 1044–1050
17. Saxena PK, King J (1990) Lack of cross-resistance of imidazolinone-resistant cell lines of *Datura innoxia P Mill.* to chlorsulfuron, Plant Physiol 94, 1111–1115
18. Wittenbach VA, Rayner DR, Schloss JV (1992) Pressure points in the biosynthetic pathway for branched-chain amino acids. In: Singh BK, Flores HE, Shannon JC (eds) Biosynthesis and Molecular Regulation of Amino Acids in Plants. American Society of Plant Physiologists, Rockville, pp 69–88
19. Butler J, Siehl D (1992) The inhibitor binding pocket of acetolactate synthase from Kochia scoparia. In: Singh BK, Flores HE, Shannon JC (eds) Biosynthesis and Molecular Regulation of Amino Acids in Plants. American Society of Plant Physiologists, Rockville, p 361
20. Subramanian MV, Gallant V, Dias JM, Mireles JC (1991) Acetolactate synthase inhibitors bind to the regulatory site, Plant Physiol 96, 310–313
21. Hawkes TR, Howard JL, Pontin SE (1989) Herbicides that inhibit the biosynthesis of branched chain amino acids. In: Dodge A (ed) Herbicides and Plant Metabolism, Seminar Series 38. Cambridge University Press, Cambridge, pp 113–136

22. Hawkes TR (1989) Studies of herbicides which inhibit branched chain amino acid biosynthesis. In: Copping LG, Dalziel J, Dodge AD (eds) Prospects for Amino Acid Biosynthesis Inhibitors in Crop Protection and Pharmaceutical Chemistry. The Lavenham Press Limited, Lavenham, pp 131–138
23. Muhitch MJ, Shaner DL, Stidham MA (1987) Imidazolinones and acetohydroxyacid synthase from plants. Properties of the enzyme from maize suspension culture cells and evidence for the binding of imazapyr to acetohydroxyacid synthase in vivo, Plant Physiol 83, 451–456
24. Shaner DL, Singh BK, Stidham MA (1990) Interaction of imidazolinones with plant acetohydroxyacid synthase: evidence for in vivo binding and competition with sulfometuron methyl, J Agric Food Chem 38, 1279–1282
25. Shaner DL, Singh BK (1991) Imidazolinone-induced loss of acetohydroxyacid synthase activity in maize is not due to the enzyme degradation, Plant Physiol 97, 1339–1341
26. Durner J, Gailus V, Böger P (1991) New aspects on inhibition of plant acetolactate synthase by chlorsulfuron and imazaquin, Plant Physiol 95, 1144–1149
27. Abell LM, Schloss JV (1991) Oxygenase side reactions of acetolactate synthase and other carbanion-forming enzymes, Biochemistry 30: 7883–7887
28. Bertagnolli BL, Hager LP (1993) Role of flavin in acetoin production by two bacterial pyruvate oxidases, Arch Biochem Biophys 300: 364–371
29. Tse JM-T, Schloss JV (1993) The oxygenase reaction of acetolactate synthase, Biochemistry, 32: 10398–10403
30. Abell LM, Schloss JV (1993) Use of the oxygenase activity of acetolactate synthase for herbicide detection. US Patent 5,206,135, 28 December 1990
31. Schloss JV, Van Dyk DE, Vasta JF, Kutny RM (1985) Purification and properties of *Salmonella typhimurium* acetolactate synthase isozyme II from *Escherichia coli* HB101/pDU9, Biochemistry 24, 4952–4959
32. Hartman FC, Harpel MR (1993) Chemical and genetic probes of the active site of D-ribulose 1,5-bisphosphate carboxylase/oxygenase: a retrospective based on the three dimensional structure, Advances Enzymol 67, 1–76
33. Lorimer GH, Andrews TJ, Tolbert NE (1973) Ribulose diphosphate oxygenase. II. Further proof of reaction products and mechanism of action, Biochemistry 12, 18–23
34. Christeller JT (1981) The effect of bivalent cations on ribulose diphosphate carboxylase/oxygenase, Biochem J 193: 839–844
35. Mogel SN, McFadden BA (1990) Chemiluminescence of the Mn^{2+}-activated ribulose-1,5-bisphosphate oxygenase reaction: evidence for singlet oxygen production, Biochemistry 29, 8333–8337
36. Gailus V (1990) Vergleichende Untersuchungen an der Acetolactatsynthase aus höheren Pflanzen, Diplom Thesis, Universität Konstanz, Constance, Germany
37. McCord JM, Fridovich I (1969) Superoxide dismutase an enzymic function for erythrocuprein (hemocuprein), J Biol Chem 244, 6049–6055
38. Corey EJ, Mehrotra MM, Khan AU (1987) Water induced dismutation of superoxide anion generates singlet molecular oxygen, Biochem Biophys Res Commun 145, 842–846
39. Kanofsky J (1983) Singlet oxygen production by lactoperoxidase, J Biol Chem 258, 5991–5993
40. LaRossa RA, Van Dyk TK, Smulski DR (1990) A need for metabolic insulation: lessons from sulfonylurea genetics. In: Barak Z, Chipman DM, Schloss JV (eds) Biosynthesis of Branched Chain Amino Acids. VCH, Weinheim, pp 109–121
41. Van Dyk TK, LaRossa RA (1990) Prevention of endogenous 2-ketobutyrate toxicity in *Salmonella typhimurium*. In: Barak Z, Chipman DM, Schloss JV (eds) Biosynthesis of Branched Chain Amino Acids. VCH, Weinheim, pp 123–130
42. LaRossa RA, Van Dyk TK (1987) Metabolic mayhem caused by 2-ketoacid imbalances, Bioessays 7, 125–130

43. LaRossa RA, Van Dyk TK, Smulski DR (1987) Toxic accumulation of α-ketobutyrate caused by inhibition of the branched-chain amino acid biosynthetic enzyme acetolactate synthase in *Salmonella typhimurium*, J Bacteriol 169, 1372–1378
44. Van Dyk TK, LaRossa RA (1986) Sensitivity of a Salmonella typhimurium AspC mutant to sulfometuron methyl, a potent inhibitor of acetolactate synthase II, J Bacteriol 165, 386–392
45. Van Dyk TK, LaRossa RA (1987) Involvement of ack-pta operon products in α-ketobutyrate metabolism by *Salmonella typhimurium*, Mol Gen Genet 207, 435–440
46. Van Dyk TK, Smulski DR, Chang YY (1987) Pleiotropic effects of poxA regulatory mutations of *Escherichia coli* and *Salmonella typhimurium*, mutations conferring sulfometuron methyl and α-ketobutyrate hypersensitivity, J Bacteriol 169, 4540–4546
47. Epelbaum S, Landstein D, Arad S (M), Barak Z, Chipman DM, LaRossa RA, Van Dyk TK (1992) Is the inhibitory effect of the herbicide sulfometuron methyl due to 2-ketobutyrate accumulation. In: Singh BK, Flores HE, Shannon JC (eds) Biosynthesis and Molecular Regulation of Amino Acids in Plants. American Society of Plant Physiologists, Rockville, pp 352–353
48. Singh BK, Shaner DL (1992) Carbon flow through branched-chain amino acid biosynthetic pathway: lessons from acetohydroxy acid synthase inhibitors. In: Singh BK, Flores HE, Shannon JC (eds) Biosynthesis and Molecular Regulation of Amino Acids in Plants. American Society of Plant Physiologists, Rockville, pp 354–357

Recent Developments in the Chemistry of Sulfonylurea Herbicides

S. K. Gee and J. V. Hay

DuPont Agricultural Products, Stine-Haskell Research Center, Newark, Delaware 19711, USA

1 Introduction

Since their discovery by Levitt in the early 1970,s the sulfonylurea herbicides have been a major research program for many crop protection chemical companies. The importance of this class of herbicide, with their unprecedented levels of activity and the breadth of structural types, required the development of new synthetic methods in several chemical areas to enable the researcher to explore this class of herbicide. An in-depth review of the chemistry that has been developed for the synthesis of sulfonylureas is well beyond the scope of this manuscript. Rather, in this review we will attempt to summarize and highlight the recent developments in the chemistry of the sulfonylurea herbicides. We will cover newer and unusual synthetic methods that has been developed to prepare a variety of sulfonamides, sulfonylureas with modified bridges, and sulfonylureas containing the sulfonylurea bridge fused to the heterocyclic portion of the molecule.

2 Synthesis of Sulfonamides

The sulfonamide portion of the sulfonylurea molecule allows the greatest structural diversity of this class of herbicide. New synthetic methods for the preparation of sulfonamides have been developed, or established methods modified, to enable agricultural chemists to explore the breadth of herbicidally active sulfonylureas. These synthetic methods fall into two broad categories – (1) introduction of the desired *ortho* substituent into an aromatic system containing a sulfonamide group or an appropriate precursor, and (2) introduction of a sulfonamido group regiospecifically into the *ortho* position of the molecule that contains the desired *ortho* substituent (Fig. 1). Representative examples of each

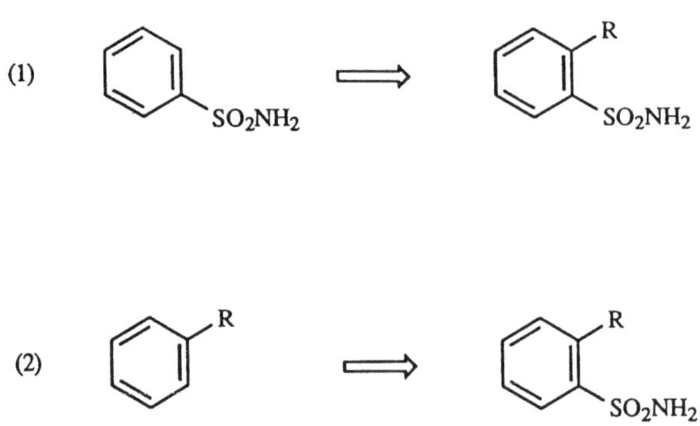

Fig. 1. General synthetic methods for preparing *ortho*-substituted sulfonamides

of these synthetic approaches, as well as several miscellaneous methods, e.g. conversion of an existing *ortho* substituent into another, will be discussed.

2.1 Introduction of an *ortho*-Substituent

Introduction of a substituent regiospecifically into the *ortho* position of an aromatic or heteroaromatic sulfonamide may be achieved by several methods. The desired substituent may be introduced into the *ortho* position of a sulfonamide by directed metallation and subsequent trapping with the appropriate electrophile, by nucleophilic displacement of a nucleofugic *ortho* substituent, or by metal mediated cross coupling reactions.

The most common method reported in the patent literature for introducing a substituent regiospecifically into the *ortho* position of an aromatic sulfonamide is directed metallation (Fig. 2). On treatment with two equivalents of an organolithium reagent, a suitably protected sulfonamide or a sulfonamide precursor, e.g. sulfonic acid, **1** directs metallation regiospecifically into the *ortho* position. Trapping of the resulting dilithiated intermediate **2** with an electrophile provides access to a wide variety of *ortho*-substituted sulfonamides **3**. Sulfonamides have been used most commonly as the metallation substrate as they are readily available and easily handled.

The most commonly used protecting group for the sulfonamide is the *t*-butyl group, based on the work of Lombardino [1] in the early 1970s. *N-tert*-butyl sulfonamides **4** are easily prepared, and after introduction of the *ortho* substituent, the *t*-butyl group is easily removed by treatment with acid, e.g. trifluoroacetic acid (TFA) to provide the free sulfonamide **6** (Fig. 3). This method has been applied to the synthesis of both aromatic and heteroaromatic sulfonamides and has provided access to otherwise inaccessible sulfonamides. French workers [2] have used this procedure to prepare *o*-trialkylsilyl-substituted benzene, thiophene and pyridine sulfonamides. For example, the *o*-triethylsilylbenzenesulfonamide **7** and the 3-trimethylsilylthiophene-2-sulfonamide **8** were obtained in good yield upon trapping the respective dilithiated intermediates with the appropriate trialkylchlorosilane (Fig. 4).

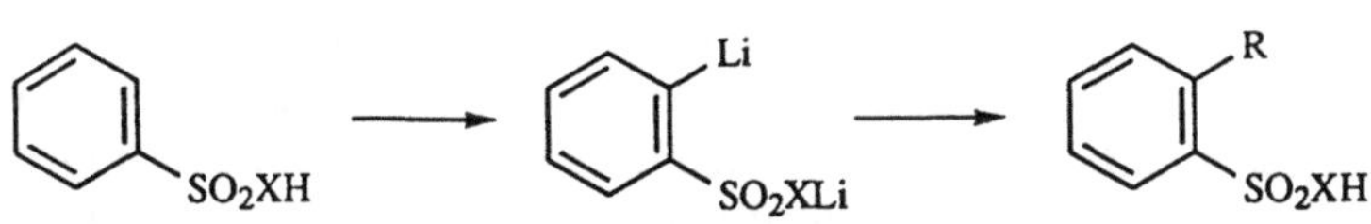

1 **2** **3**

X = NR',O

Fig. 2. Directed metallation of aromatic sulfonamides and sulfonic acids

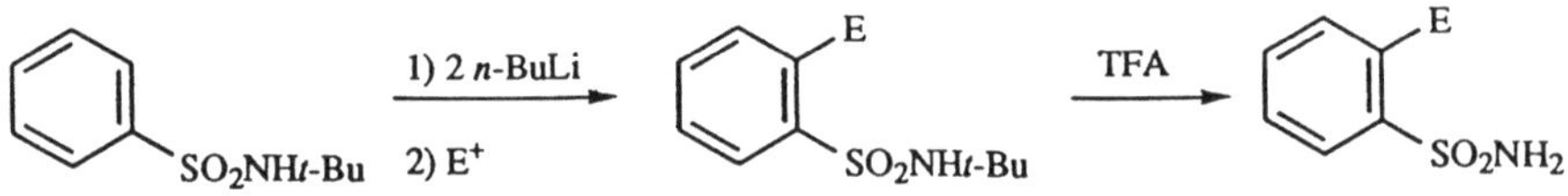

4 5 6

Fig. 3. Synthesis of *ortho*-substituted sulfonamides from *N-tert*-butyl protected sulfonamides by metallation and electrophilic trapping

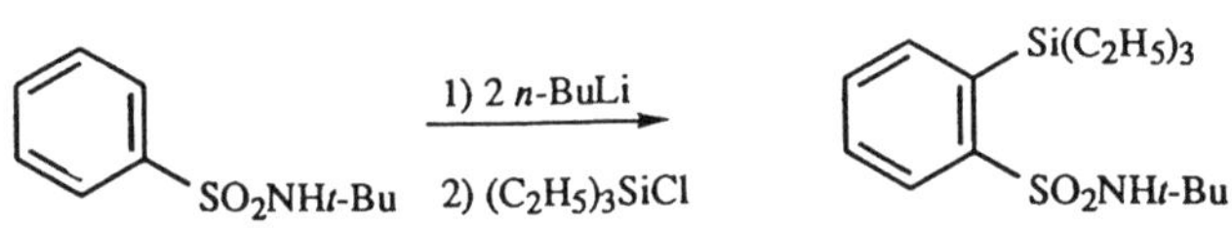

4 7

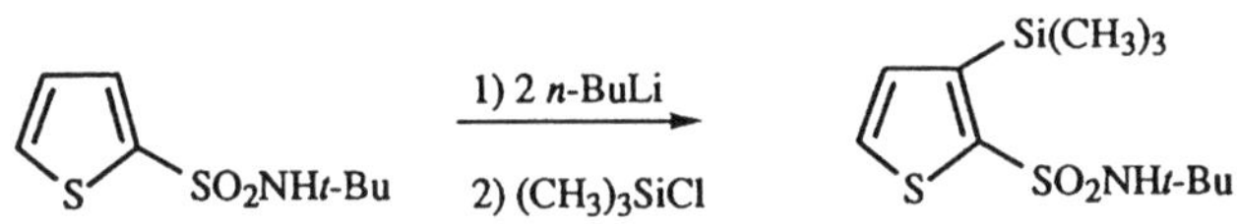

8

Fig. 4. Preparation of *ortho*-trialkylsilylbenzene and thiophene sulfonamides

Some 2,6-disubstituted benzene sulfonamides may be prepared by this method in a "one pot" reaction analogous to the method shown in Fig. 3 [3]. After trapping the dilithiated sulfonamide **4** with an electrophile, an additional equivalent of *n*-butyl lithium is added. After allowing sufficient time for metallation to occur at the vacant *ortho* position, a second electrophile (either the same or different than the first) is added to provide the 2,6-disubstituted benzene sulfonamides.

Japanese workers have used fluorinated olefins as electrophiles in similar reactions to prepare both fluoroalkyl and fluoroalkenyl substituted heteroaromatic sulfonamides (Fig. 5) [4]. Trapping of the dilithiated intermediate derived from thiophene **9** with chlorotrifluoroethylene gave sulfonamide **10**. Treatment of **10** with potassium fluoride effects both desilylation and addition of fluoride to the olefinic group to provide the tetrafluoroethylsulfonamide **11**. The same research group prepared the pentafluoropropenyl pyridine sulfonamides **13** and **15** by trapping the lithiated intermediates derived from sulfonamides **12** and **14** (Fig. 6) [5]. A variety of other substituted thiophene, pyrazole, furan, pyridine and pyrrole sulfonamides may be prepared by these methods [6–9].

This method of preparing *o*-substituted sulfonamides normally provides good yields of desired product for most electrophiles. However, low yields are often obtained when the dilithiated sulfonamide is trapped with very reactive electrophiles such as benzylic and allylic halides. In these cases, conversion of

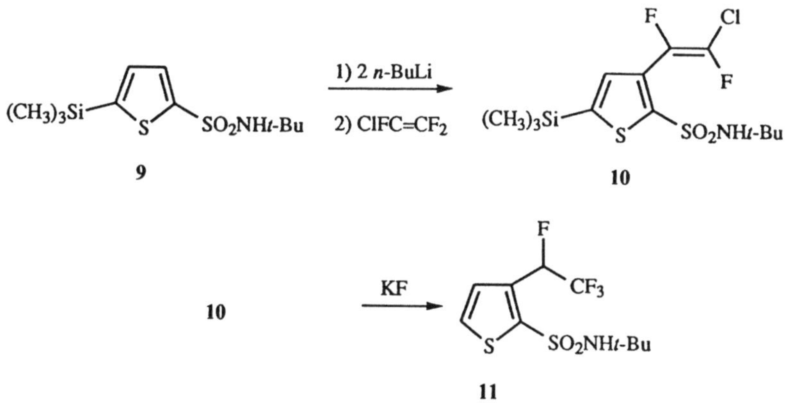

Fig. 5. Preparation of 3-tetrafluoroethyl-2-thiophene sulfonamides

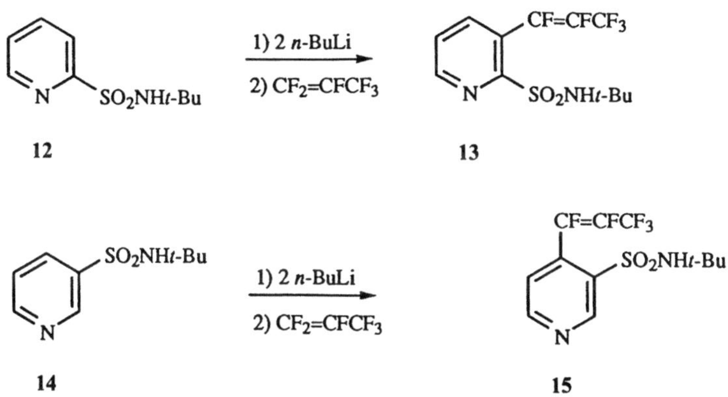

Fig. 6. Preparation of pentafluoropropenylpyridine sulfonamides via dilithiated pyridine sulfonamides

the organolithium reagent to an organozinc or organomagnesium species enables good yields of *o*-benzyl or *o*-allyl substituted sulfonamides. For example, treatment of the dilithiated intermediate derived from **1** with anhydrous zinc chloride gives the organozinc species **16**, subsequent trapping with allyl bromide and deprotection provides a good yield of *o*-allylbenzene sulfonamide **17** (Fig. 7) [10].

Reactions where an acid sensitive *ortho* group, or other labile group, is produced upon trapping of the organometallic species, a sulfonamide protecting group other than *N-tert*-butyl group must be used. In these cases, the *N-tert*-butyldimethylsilyl group is the protecting group of choice [11]. The silylated sulfonamide **18** is easily prepared, and after trapping of the *ortho* lithiated sulfonamide with the desired electrophile, the silyl protecting group is readily

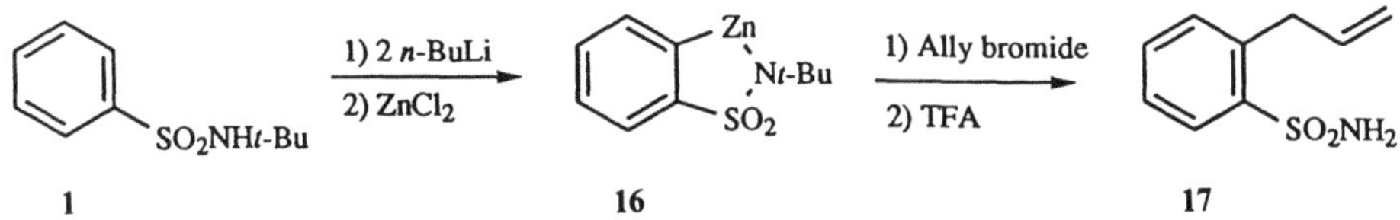

Fig. 7. Preparation of *o*-allylbenzene sulfonamide through an organozinc intermediate

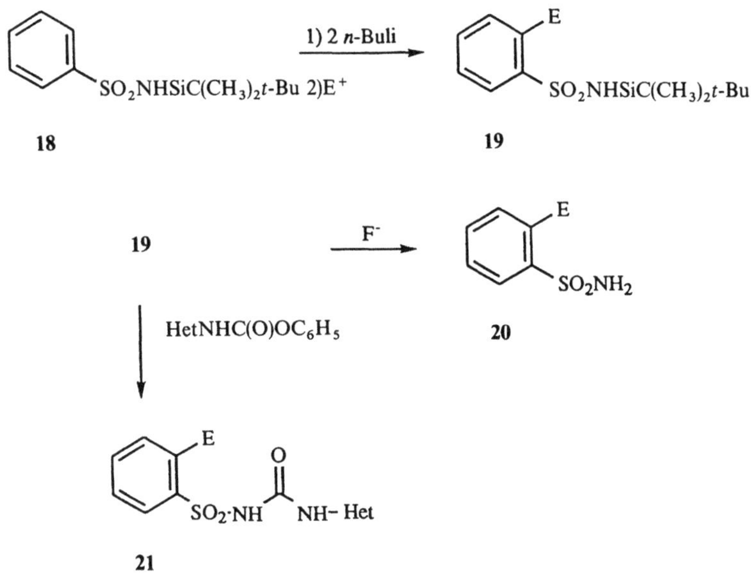

Fig. 8. Use of *N-tert*-butyldimethylsilyl protecting group for acid sensitive groups produced from lithiated sulfonamide intermediates

removed from **19** with fluoride ion to produce the free sulfonamide **20** with the acid sensitive *ortho* group intact. Alternatively, the silyl protected sulfonamide **19** can be converted directly to a sulfonylurea **21** by reaction with the appropriate heterocyclic phenyl carbamate in the presence of fluoride ion (Fig. 8).

An example of this method is shown in Fig. 9. Trapping of the dilithiated intermediate derived from **18** with cyclopropylcarboxaldehyde gives the silyl protected 2-(cyclopropyl(hydroxy)methyl)benzene sulfonamide **22**. In this instance **22** was converted directly into a sulfonylurea as in Fig. 8, but could as easily be converted to the free sulfonamide with fluoride anion.

Although the use of an aromatic sulfonic acids gives comparable yields of product as a protected sulfonamide, this method has received only limited attention. This method suffers in that conversion of the resulting sulfonic acid to the desired sulfonamide requires two steps (chlorination and amination) versus only one (deprotection) for the N-protected sulfonamide. Although lengthier

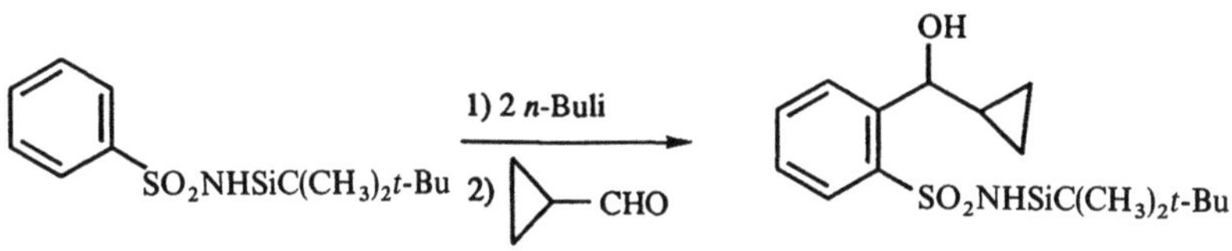

18 22

Fig. 9. Preparation of 2-(cyclopropyl(hydroxymethyl)benzenesulfonamide via silyl protected sulfonamide

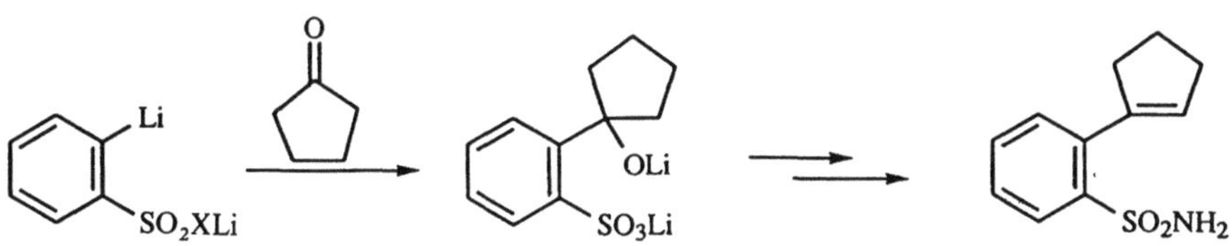

2 23 24

Fig. 10. Preparation of 2-(cycloalkenyl)benzene sulfonamides via dilithiated sulfonic acids

than the protected sulfonamide route, Chen and Schwing [12] utilized this method to prepare a series of 2-(cycloalkenyl)benzenesulfonamides in good to moderate yield (Fig. 10). Treatment of the dilithiated intermediate **2** where X = O, with cyclopentanone gives the lithium alkoxide **23**, which on reaction with thionyl chloride followed by ammonia, gave 2-(1-cyclopentenyl)benzene sulfonamide **24**.

Introduction of a substituent by a nucleophilic process into the *ortho* position of an aromatic or heteroaromatic sulfonamide has not been widely exploited by agricultural chemists. Substrates for this synthetic method are normally sulfonamides, either unprotected or protected, substituted at the *ortho* position with a nucleofugic group, such as a halogen. The chlorine group of 2-chlorobenzene sulfonamide **25** can be displaced with nucleophiles such as mercaptides, aliphatic amines and selenides (Fig. 11) [13, 14]. Similarly, Föry prepared pyridine sulfonamide **27** by displacement of fluoride in **26** with 2-fluoroethoxide ion [15]. A variety of substituted thiophene, pyrazole and isothiazole sulfonamides can be prepared by similar methods [6, 16].

A third method of introducing a substituent directly into the *ortho* position of an aromatic or heteroaromatic sulfonamide is by transition metal mediated processes. Shapiro found that thiophene sulfonamide **28** could be coupled with both the boronic acid **29** and the trimethyltin derivative **31** to give the arylated thiophene sulfonamides **30** and **32** respectively in the presence of the palladium catalyst $Pd[P(C_6H_5)]_4$ (Fig. 12) [17].

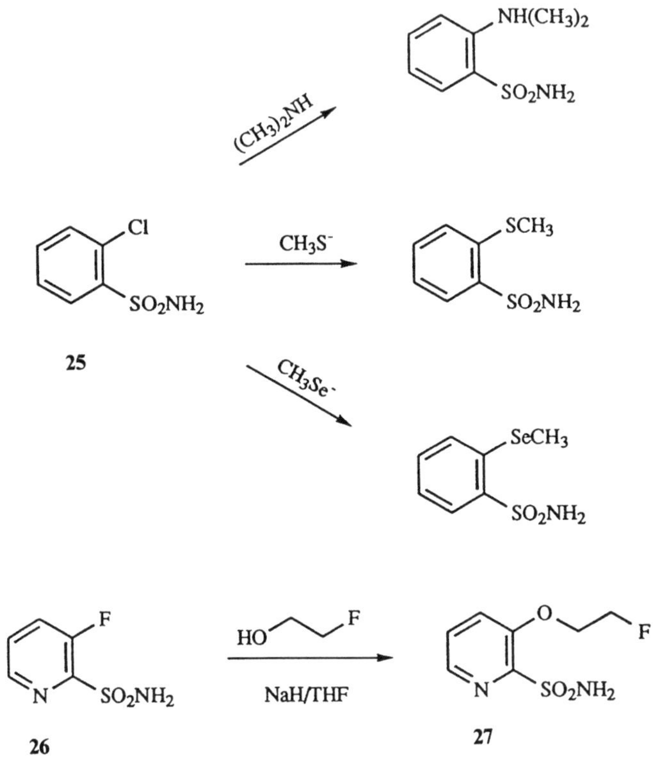

Fig. 11. Preparation of *ortho*-substituted sulfonamides by aromatic nucleophilic substitution

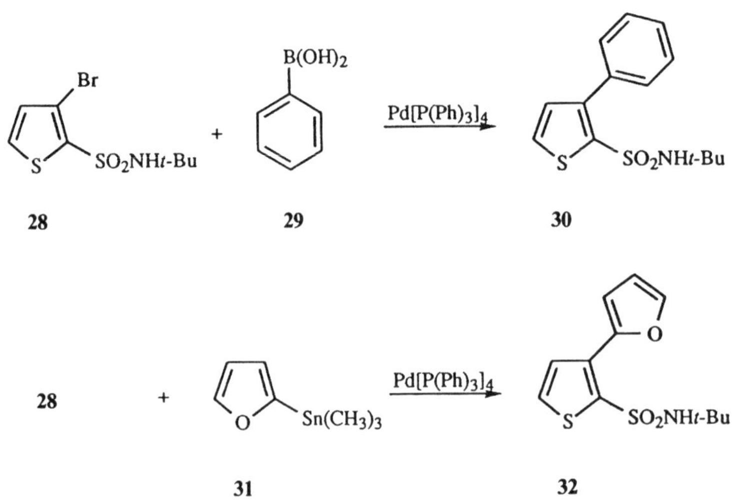

Fig. 12. Palladium catalyzed arylation and heteroarylation of thiophene sulfonamides

Similarly, *ortho* acetylenic groups can be introduced by a palladium catalyzed coupling reaction. For example, 2-(1-propynyl)benzene sulfonamide **34** can be prepared from the iodosulfonamide **33** and propyne using $PdCl_2(C_6H_5)_3$ as shown in Fig. 13 [18].

Biphenyl sulfonamides, e.g. **36**, can be prepared by reaction of an aryl iodide and the aryl copper reagent **35** (Fig. 14) [19]. Likewise, the organozinc reagent **16** undergoes coupling with 3-bromopyridine in the presence of a palladium catalyst to give sulfonamide **37** (Fig. 15) [20].

Diazonium salts derived from 2-aminobenzene sulfonic acid have also been used as substrates for metal mediated coupling reactions to produce ultimately 2-alkenylbenzene sulfonamides [21, 22]. Treatment of the diazonium tetrafluoroborate **38** with 3,3,3-trifluoropropene or allyl acetate in the presence of bis-(benzylideneacetone)palladium(0) gives the alkenyl sulfonamides **39** and **40** respectively (Fig. 16).

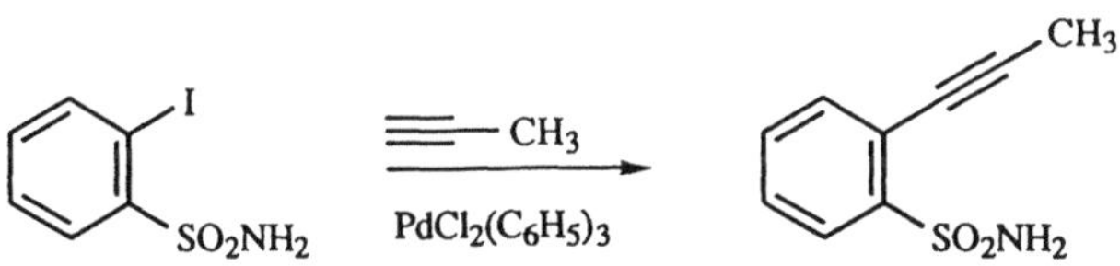

33 **34**

Fig. 13. Palladium catalyzed synthesis of *ortho*-acetylenic sulfonamides

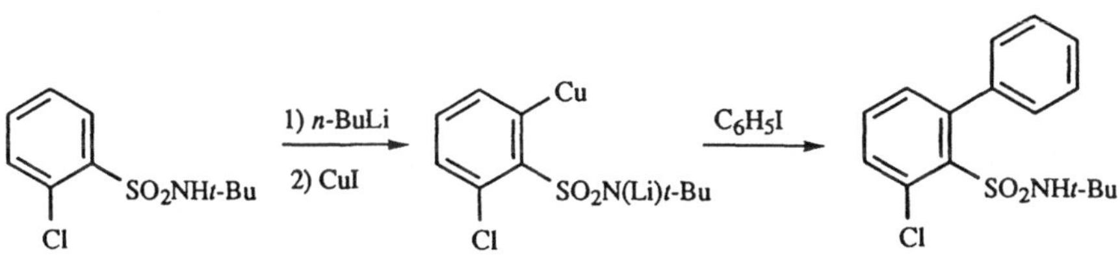

35 **36**

Fig. 14. Phenylation of arylcopper reagents

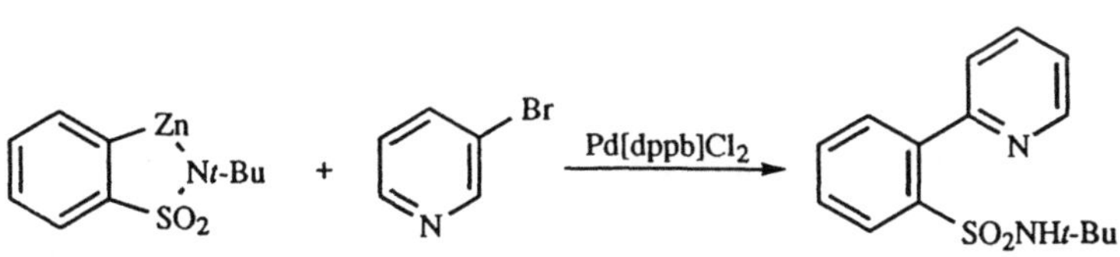

16 **37**

Fig. 15. Palladium catalyzed heteroarylation of an organozinc reagent

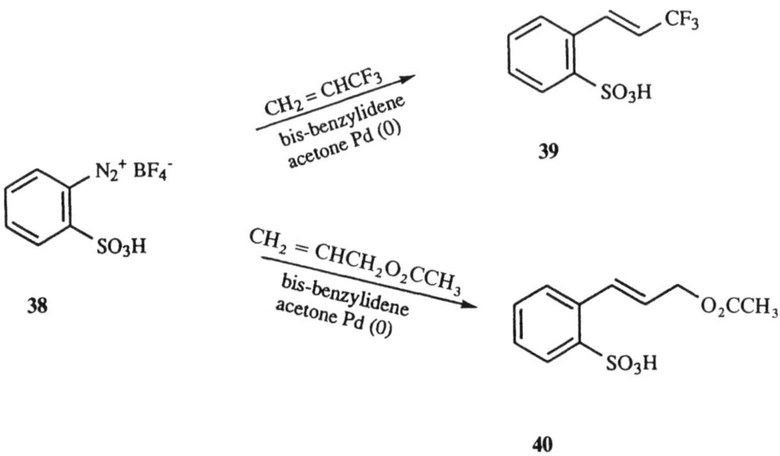

Fig. 16. Palladium catalyzed coupling of alkenes with diazonium salts

2.2 Introduction of an *ortho*-Sulfonamide Group

Relatively few new methods for regiospecifically introducing a sulfonamide group into the *ortho* position have appeared in the literature since this topic was reviewed in 1990 [23]. Rather, most research has focused on developing improvements for, or modifying existing synthetic methods. One such modification, reported by Wolf [24], is the conversion of the sulfinate salt **41** directly to the sulfonamide **42** with chloramine (Fig. 17). This method gives higher yields of sulfonamide than the conventional method of converting the sulfinate salt to the sulfonyl chloride followed by amination.

Similarly, direct conversion of the thiophene sulfinate salt **43** to the sulfonamide **44** with hydroxylamine-*O*-sulfonic acid provides higher yields than the two step method (Fig. 18) [6]. This method may be used for heteroaromatic systems (pyrazoles, pyridines, pyrroles) or simple aromatic systems. Use of sulfuryl chloride fluoride in place of sulfuryl chloride during the conversion of bromostryene **45** to sulfonamide **46** gives a cleaner product and higher yield than the latter reagent (Fig 19) [11]. The yield improvement is believed to be due to the greater hydrolytic stability of sulfonyl fluorides relative to sulfonyl chlorides.

2.3 Transformation of *ortho*-Substituents

Conversion of an *ortho*-substituent on an arylsulfonamide into a different *ortho*-substituent has received much attention from agricultural chemists. This transformation has enabled researchers to quickly explore structure-activity relation-

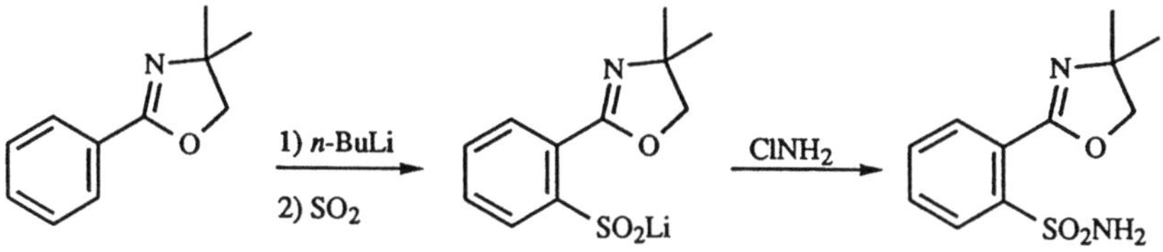

41 **42**

Fig. 17. Conversion of a sulfinate salt to a sulfonamide with chloroamine

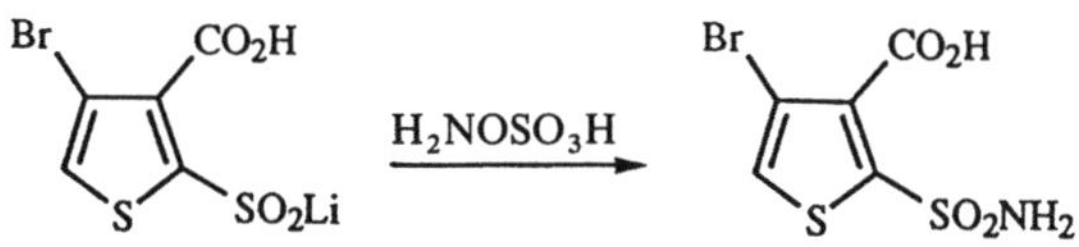

43 **44**

Fig. 18. Conversion of a sulfinate salt to a sulfonamide with hydroxylamine-*O*-sulfonic acid

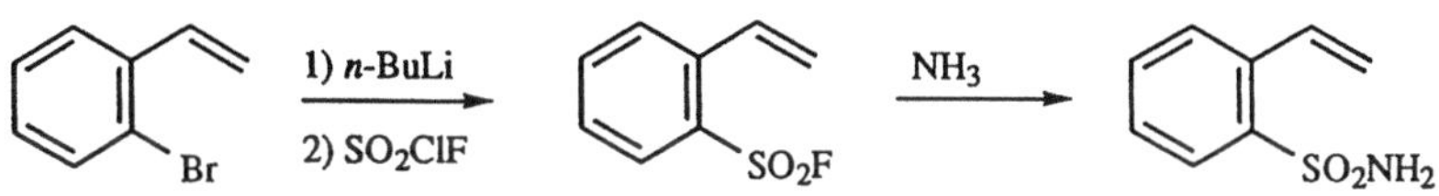

45 **46**

Fig. 19. Preparation of sulfonamides using sulfuryl chloride fluoride

ships. Meyer prepared the *ortho*-allenic ether **49** from either the propargyl ether **47** or the chloroallyl ether **48** by treatment with potassium *t*-butoxide (Fig. 20) [25].

The hemiaminal **50** has proven to be a versatile intermediate for the preparation of a variety of *ortho*-substituted sulfonamides (Fig. 21). Under Wittig conditions, treatment of **50** with ylide **51** affords the vinyl lactone **52** [26]. In the presence of $Eu(fod)_3$, **50** undergoes cycloaddition with Danishefsky's diene **53** to the dihydropyran-4-one substituted sulfonamide **54** [27]. Other transformations of hemiaminal **50** have been reported [23, 28]. Analogously, keto (aldehydo) thiophene **55** can be transformed into the oxazole **56** with tosylmethylisocyanide, to the oxime **57** with methoxyamine, to the difluoromethylthiophene **58** with DAST, or to the ketal **59** with ethylene glycol (Fig. 22) [6]. Rorer [29] and Hartzell [30] converted esters **60** and **62** into the *ortho*-thiadiazole **61** or oxadiazole **63** through an intermediate hydrazide (Fig. 23). Other *ortho*-heterocyclic substituted sulfonamides have been prepared in a similar fashion from an appropriate *ortho* precursor [24, 31, 32]. Thompson found that lactone **66** could be prepared from either the benzothiazine **64** or phenyl acetate **65** by metallation and subsequent trapping with ethylene oxide

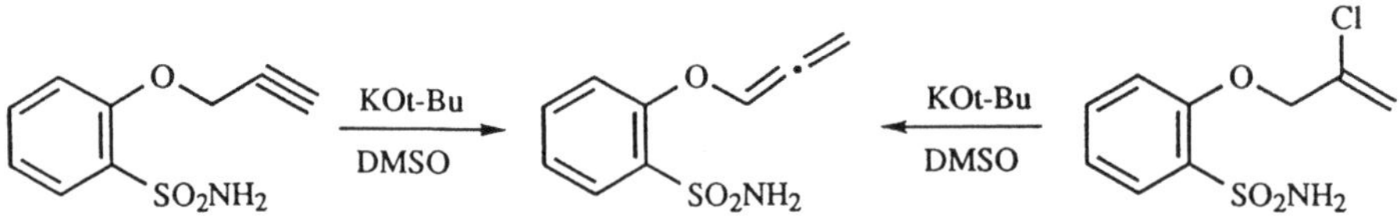

47 **49** **48**

Fig. 20. Allenic ether formation from propargylic or allylic ethers

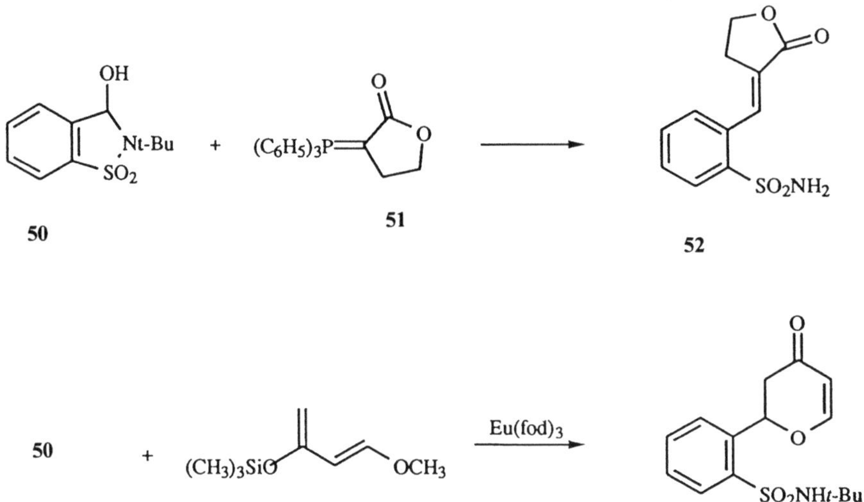

50 **51** **52** **53** **54**

Fig. 21. Transformations of *N*-sulfonyl hemiaminal into cyclic *ortho*-groups

(Fig. 24) [26]. Propylene oxide also afforded the corresponding methylated lactone, while other epoxides failed to react with the metallated intermediate derived from **64**. *o*-Keto sulfonamides have been obtained from the corresponding acids [33], and epoxides and cyclopropanes from an *o*-vinyl sulfonamide [11].

2.4 Miscellaneous Methods of Sulfonamide Synthesis

An exhaustive review of all methods utilized for the synthesis of sulfonamides is beyond the scope of this review. Only selected methods are highlighted. Petersen reported that dihydroindole and tetrahydroquinoline sulfonamides can be readily prepared by the method shown in Fig. 25 [34]. Treatment of dihydroindole **67** with chlorosulfonylisocynate followed by aluminium chloride

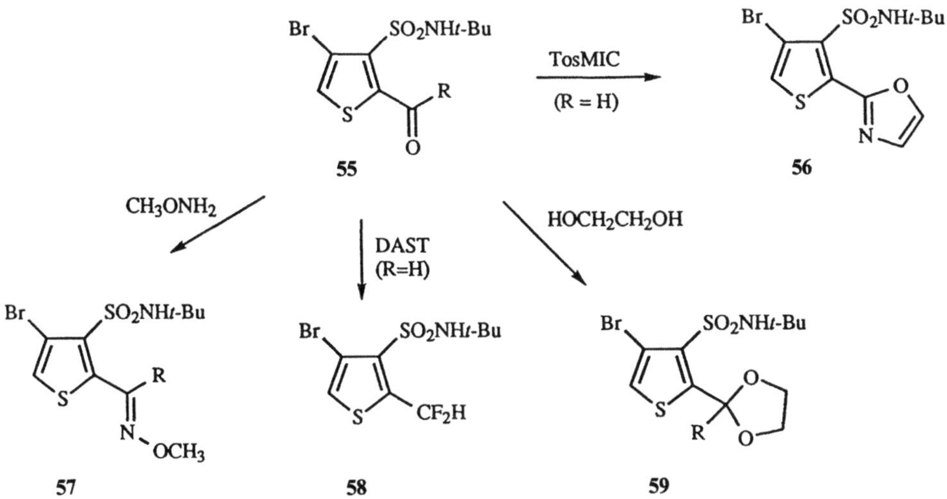

Fig. 22. Conversion of thiophene ketones and aldehydes to other *ortho*-groups

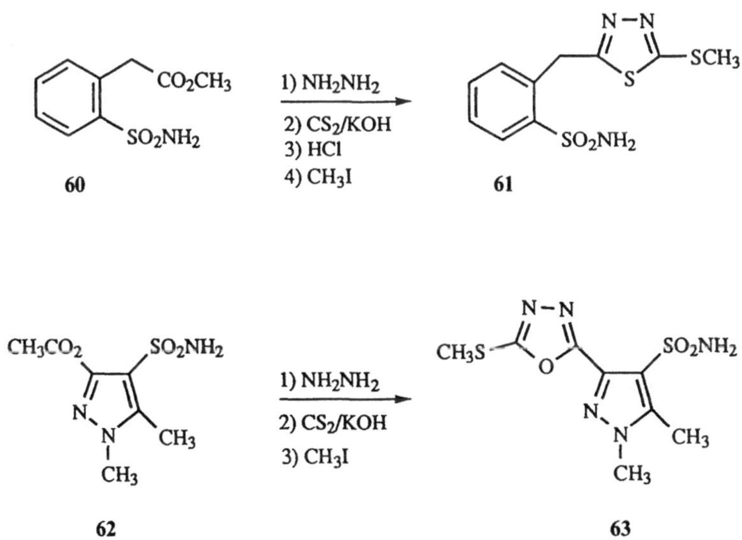

Fig. 23. Synthesis of sulfonamides bearing *ortho*-heterocyclic groups from esters

in nitroethane gives the tricyclic sultam **68**. Acidic hydrolysis of **68** gives the dihydroindole sulfonamide **69**.

Föry reported an unusual reaction that produces 1,2-benzoxathiine sulfonamides (Fig. 26) [35]. Treatment of dihydrobenzofuran **70** with chlorosulfonic acid not only effects chlorosulfonation, but also ring expansion to give after amination the 1,2-benzoxathiine ring system, **71**. Hydrogenolysis of **71** gave the dihydro-1,2-benzoxathiine sulfonamide **72**.

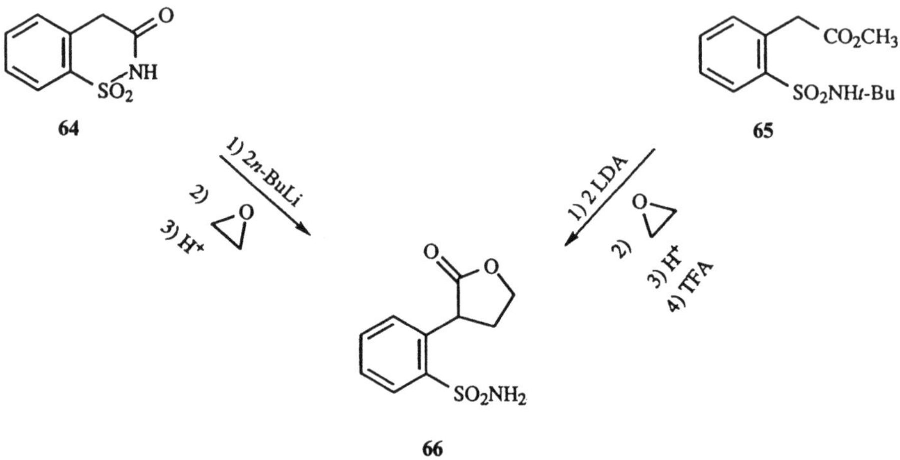

Fig. 24. Preparation of 2-(tetrahydrofuran-2-one-3yl)benzene sulfonamide

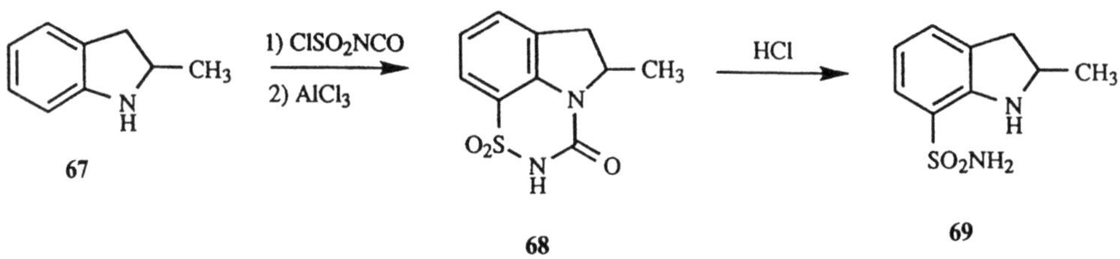

Fig. 25. Preparation of dihydroindole sulfonamides

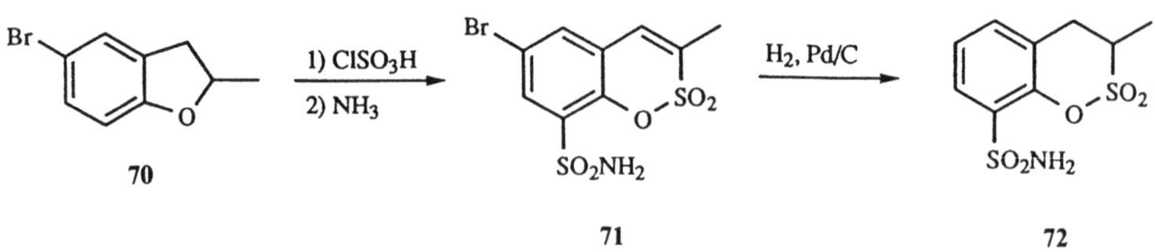

Fig. 26. Synthesis of 1,2-benzoxathiine sulfonamides

o-Sulfobenzoic anhydride **73** has often been used to prepare sulfonamides bearing *ortho* ester groups [36]. Chen and Schwing found that *o*-alkenyl substituted sulfonamides may also be prepared from **73** (Fig. 27) [12]. Treatment of **73** with excess methyl lithium gives the cyclic sulfonate **74**. Upon treatment of **74** with potassium *t*-butoxide, elimination occurs giving sulfonic acid salt **75**. Conversion of **75** to 2-isopropenyl sulfonamide **76** is readily effected

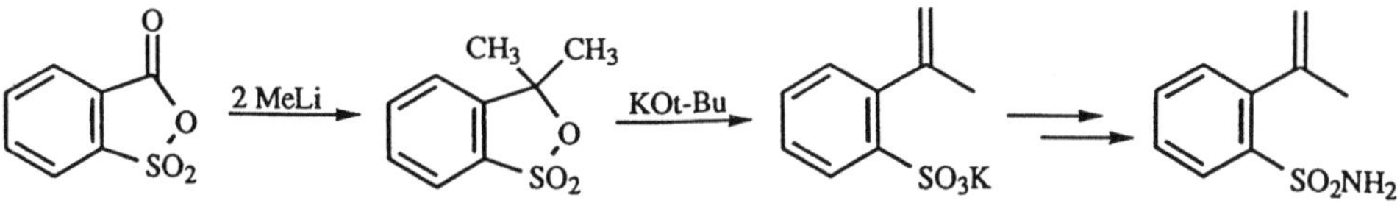

73 74 75 76

Fig. 27. Synthesis of sulfonamides from *o*-sulfobenzoic anhydride

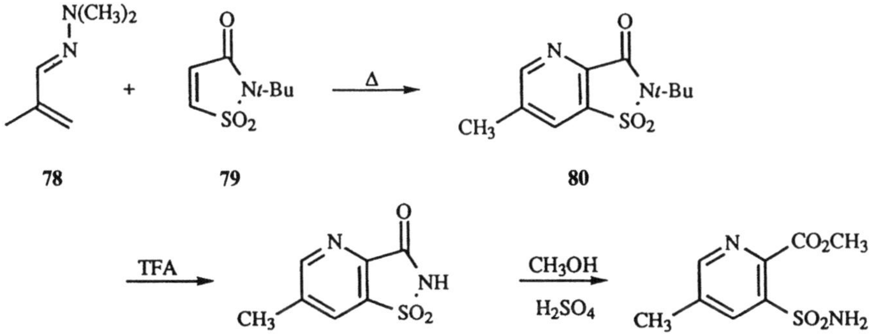

78 79 80

81 77

Fig. 28. Preparation of 2-alkoxycarbonyl-3-pyridine sulfonamides

by chlorination and then amination. A variety of 5-substituted-2-alkoxycarbonylpyridine-3-sulfonamides, e.g. **77**, may be prepared by the route shown in Fig. 28 [36]. Thermal cycloaddition of azabutadiene **78** with isothiazoline dioxide **79** gives the *N-tert*-butylazasaccharin **80**. Nitrogen deprotection with TFA, followed by methanolysis of **81** gave the methyl ester **77**.

3 New Methods for the Synthesis of Sulfonylureas

Several previously published reviews [23, 37] have described in detail the synthesis of sulfonylureas and their intermediate sulfonamides. The following section will detail newer methods that have recently published as well as those methods not covered in the preceeding reviews.

3.1 Normal Bridge Sulfonylureas (Fig. 29; m,n = 0)

A classic strategy utilized in the synthesis of sulfonylureas is the reaction of a sulfonylisocyanate with a heterocyclic amine. Several of the recently reported

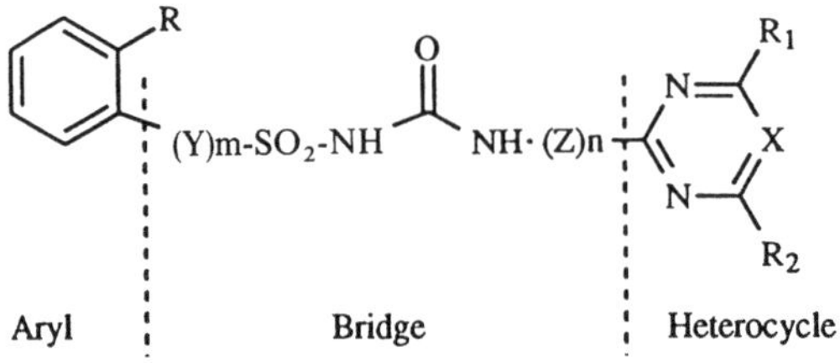

Fig. 29. Normal sulfonylurea general structure

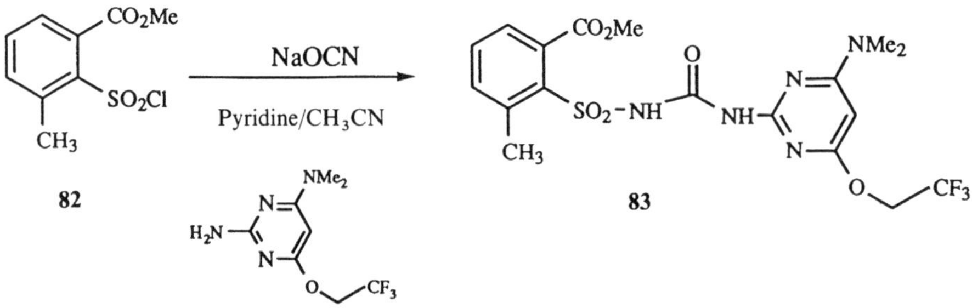

82

83

Fig. 30. Synthesis of 2,6-disubstitutedbenzenesulfonylureas

newer methods make use of this general sequence but apply new technology which allows the in situ generation of the reactive sulfonylisocyanate. The *in situ* generation of the sulfonylisocyanate has the advantage of not requiring the isolation of this reactive intermediate and also allows the reaction to be conducted in one pot. As shown in (Fig. 30), 2,6-disubstituted sulfonylurea **83** is readily prepared via the reaction of an in situ generated sulfonylisocyanate with a heterocyclic amine [38]. This sulfonylurea can be difficult to prepare by a classical method such as phenyl carbamate coupling (i.e. the coupling of an arylsulfonamide with the phenyl carbamate of the heterocyclic amine) due to the unwanted formation of saccharins. Similar conditions for the in situ generation of sulfonylisocyanate have also been reported for the synthesis of simple *ortho*-substituted benzenesulfonylureas [39, 40].

Another method that has been employed for the in situ generation of sulfonylisocyanates is the palladium catalyzed insertion of carbon monoxide into a substituted sulfonamide. When a solution of the diphenylselenylimine **84** and a heterocyclic amine is heated in the presence of a palladium catalyst, under an atmosphere of carbon monoxide, the expected sulfonylureas are obtained in excellent overall yield (Fig. 31) [41]. This same group has also reported that a variety of sulfonyl imines can be employed with equal utility in this reaction (i.e. $ArSO_2N=XR^1R^2R^3$ where X is an atom of phosphorous, sulfur, tellurium, arsenic, antimony or iodine). Besenyei and co-workers have also reported [42, 43] a palladium catalyzed carbon monoxide insertion reaction employing N-halogen-sulfonamidates ($ArSO_2NClNa$) in place of the sulfonylimines.

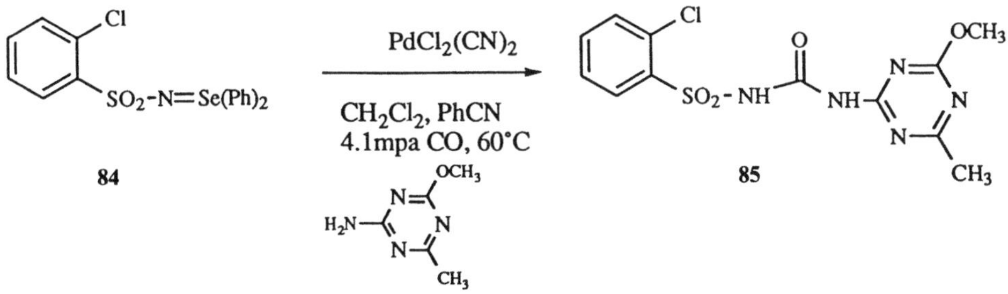

Fig. 31. Synthesis of sulfonylureas via palladium catalyzed carbon monoxide insertion

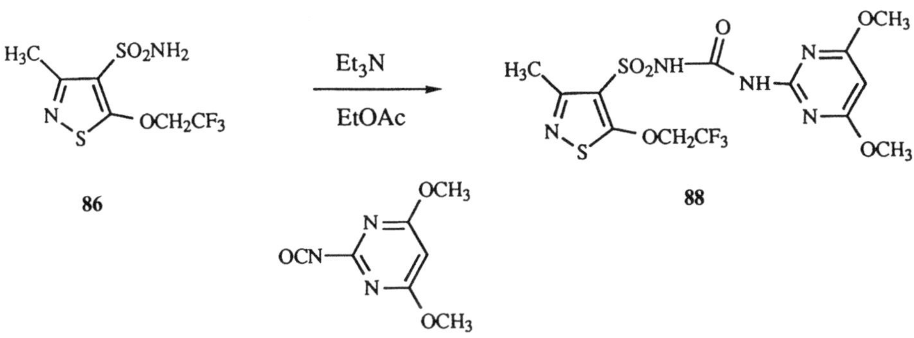

Fig. 32. Pyrimidinylisocyanate coupling reaction

Another approach that has received somewhat less attention is the reaction of sulfonamides with heterocyclic isocyanates or equivalents, such as chlorocarbamates [44, 45]. Haga and coworkers [44] have reported the preparation of the isothiazolesulfonylurea **88** via the treatment of sulfonamide **86** with the 2-pyrimidinylisocyanate in the presence of triethylamine (Fig. 32). The 2-pyrimidinylisocyanates can be prepared by the treatment of aminopyrimidines with phosgene. This reaction has been applied with equal success to both pyrimidinyl- and triazinylisocyanates. The high reactivity of the heterocyclic isocyanate allows the use of very mild reaction conditions for this coupling process. The authors also indicate that this process lends itself well to large scale synthesis.

A novel approach to the synthesis of certain specific *ortho*-substituted benzenesulfonylureas which contain either a sulfonamide or carboxamide in the *ortho* position has been reported by Diehr and co-workers [46, 47, 48]. This two step sequence involves the reaction of a heterocyclic substituted oxyguanidine **90** with either benzene-1,2-disulfonic acid dichlorides or 2-chlorosulfonylbenzoyl chlorides to produce an initial cyclic guanidine. In a second step the cyclic guanidine is cleaved via acid catalyzed hydrolysis to produce the corresponding 2-sulfonyl- or 2-carbamoylbenzenesulfonylureas (Fig. 33). This reaction sequence appears to be limited to the use of *N*-alkoxyguanidines and so limits

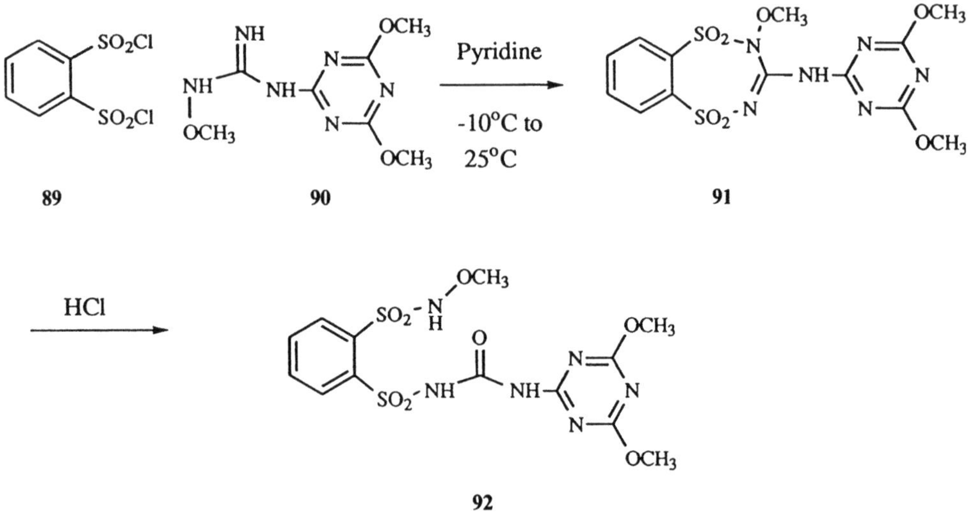

Fig. 33. Sulfonylureas via acid catalyzed hydrolysis of cyclic guanidines

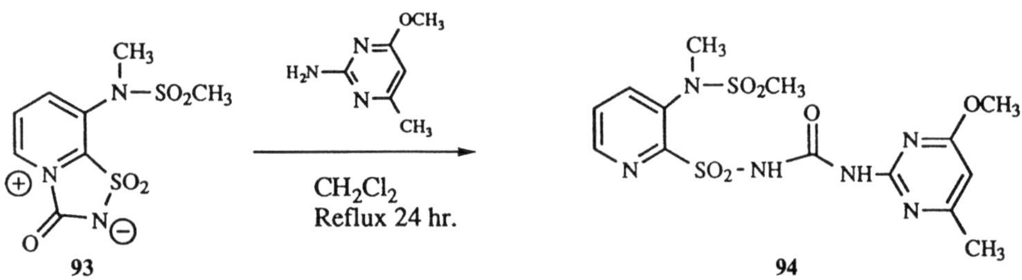

Fig. 34. Synthesis of sulfonylureas via cleavage of 1,1,3-trioxo-1,2,4-thiadiazole[4,5-a]pyridines

somewhat the functionality that can be incorporated in the *ortho* group. The cyclic guanidines also serve as intermediates in the preparation of acyclic guanidines [49], isothioureas and isoureas [50].

A interesting new process for the preparation of 2-pyridylsulfonylureas which proceeds under very mild conditions and without the need for basic catalysis has recently been reported (Fig. 34) [51]. Heating a solution of a heterocyclic amine with the thiadiazole [4,5-a]pyridine **93** in methylene chloride at reflux for several hours provides the expected sulfonylureas in near quantitative yield. This process is particularly attractive in that the product is produced as the sole component in the reaction which greatly simplifies isolation and purification. The intermediate thiadiazole[4,5-a]pyridines are produced by the treatment of *N*-pyridylsulfonyl-*N′*-alkylureas with phosgene. The overall process from the *N*-pyridylsulfonyl-*N′*-alkylurea to the final product can also be carried out as a one pot reaction.

N-Alkylbenzenesulfonylureas have been prepared from the corresponding *N*-methylcarbamoyl chlorides by treatment with a heterocyclic amine (Fig. 35) [52]. The heterocyclic amine can be either primary or secondary. The use of a secondary amine results in sulfonylureas with substitution present on both nitrogen of the sulfonylurea bridge. The corresponding sulfonylthioureas can also be prepared in an analogous manner by using the thiocarbamoyl derivative.

Certain electron rich heterocyclic sulfonylureas can be prepared by electrophilic substitution reactions with chlorosulfonylureas (Fig. 36). This reaction scheme has found utility in the preparation of 3-pyrrole- [53] and 3-indolesulfonylureas [54]. A heterocyclic amine is first treated at low temperature with chlorosulfonyl isocyanate to provide the intermediate chlorosulfonylurea **98**. This solution is then treated with an appropriate electron rich heterocycle. This reaction can also be catalyzed by Lewis acids such as $AlCl_3$. This reaction also tolerates substitution on the amino group of the heterocycle (Fig. 36, R) and sulfonylureas with an alkyl or alkoxy group on the sulfonylurea bridge have been prepared using this methodology.

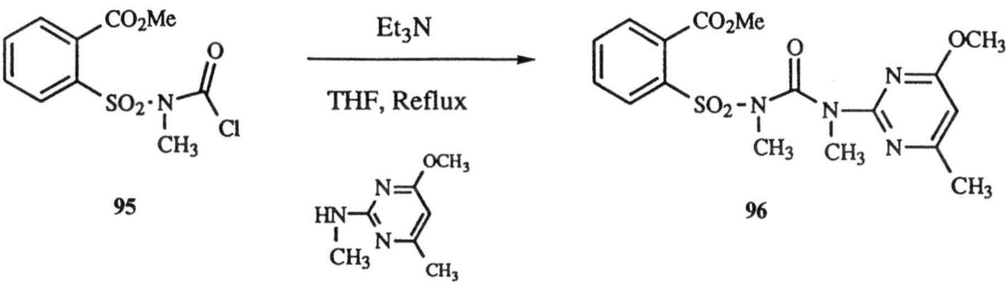

Fig. 35. Sulfonylureas from arylsulfonyl-*N*-carbamoyl chlorides

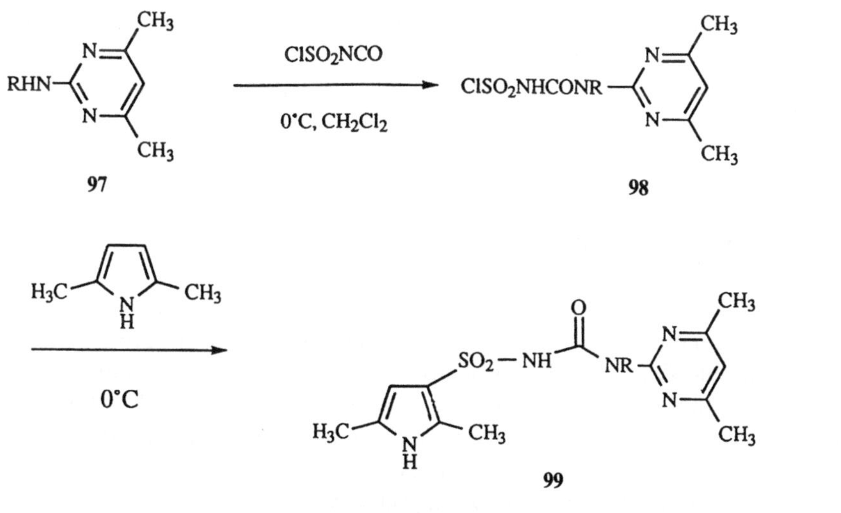

Fig. 36. Sulfonylureas via electrophilic substitution reactions of chlorosulfonylureas

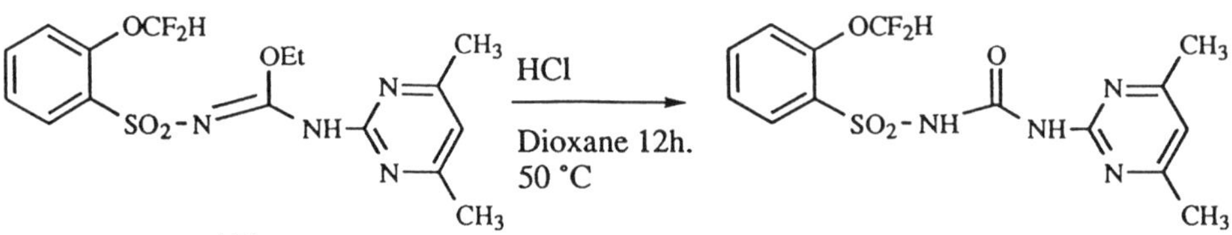

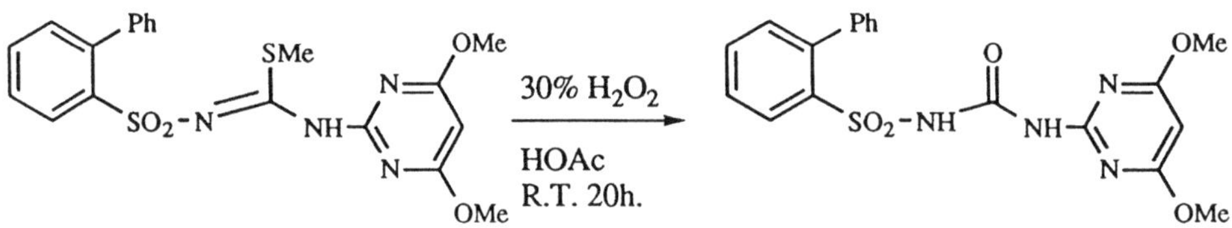

Fig. 37. Sulfonylureas from the hydrolysis of sulfonylisoureas or sulfonylisothioureas

Another approach to the synthesis of sulfonylureas that has been utilized in certain circumstances is the hydrolysis of sulfonylisoureas [55] and sulfonylisothioureas [56] (Fig. 37). For example, when the sulfonylisourea **100** is treated with a solution of HCl in dioxane the sulfonylurea **101** is produced in 97% yield. Since the sulfonylureas are produced in the absence of any other reagents this methodology is particularly well suited for the production of high purity sulfonylureas. Another method which claims to simplify isolation, to produce sulfonylureas of very high purity, and without the use of expensive catalysts such as DBU is the hydrolysis of sulfonylisothioureas [56]. Treatment of a solution of sulfonylisothiourea **102** with 30% hydrogen peroxide yields the sulfonylurea **103**. Additional oxidants that have been investigated for this sequence include MCPBA and $KMnO_4$.

3.2 Bridge Extended Sulfonylureas

In general, the high level of biological activity observed for the sulfonylurea herbicides remains even with fairly extensive modification to the gross structure of the molecule. One general class of modification that has received considerable attention is the insertion or addition of atoms into the normal sulfonylurea bridge. The vast majority of the insertions into the bridge have been additions of carbon or heteroatoms at the terminus of the bridge at either the aryl or heterocycle groups (Fig. 29, Y or Z respectively).

3.2.1 Carbon Based Bridge Extensions (Fig. 29, Y or Z = C)

By far the most common carbon based modification to the sulfonylurea bridge is insertion of a single carbon atom between the aryl portion and the sulfonyl group (Fig. 29, Y = C, m = 1). One particularly noteworthy example of this modification is the rice herbicide Londax® (bensulfuron methyl) (Fig. 38).

Almost all of the sulfonylureas of this general structural class have been prepared by displacement of a benzyl halide with thiourea followed by oxidation of the resultant isothiouronium salt **105** to the sulfonyl chloride. Conversion to the benzylsulfonamide is effected by treatment with ammonia (Fig. 39). The sulfonamides can then be carried on to the sulfonylureas by any of the previously reviewed coupling methodology [23, 37]. The displacement of the benzyl halide is usually quite facile and is conducted by heating an alcoholic solution of thioureas with the benzyl halide [57]. As an alternative to thiourea,

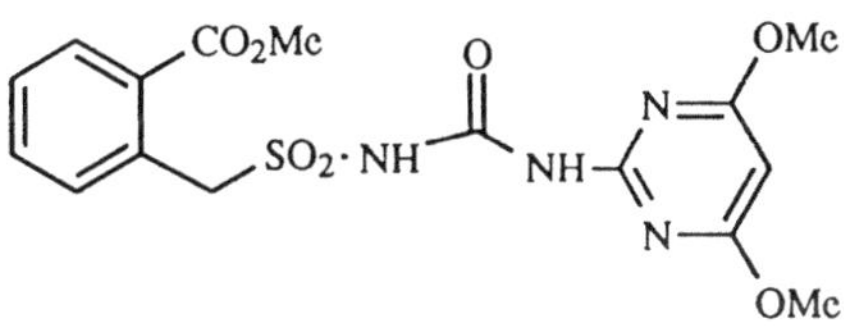

Fig. 38. Rice herbicide Londax

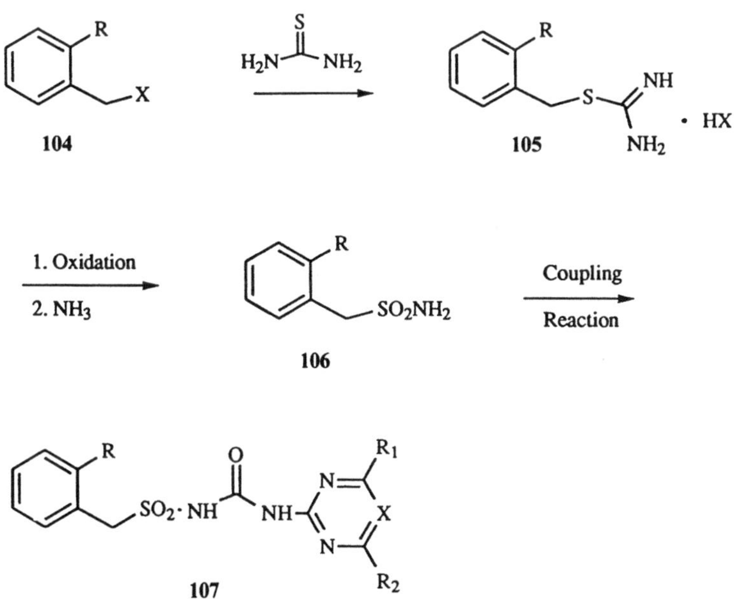

Fig. 39. Synthesis of benzylsulfonylureas via isothiouronium salts

mercaptides [58] have been employed with equal utility. The isothiuronium salt is typically oxidized to the sulfonyl chloride via treatment with chlorine [59]. Alternatively, oxidation has been accomplished with a sodium hypochlorite solution [60]. This methodology has been utilized in the preparation of simple benzylsulfonylureas [61], 2,6-disubstituted-benzylsulfonylureas [62], 2-keto-benzylsulfonylureas [63], thiophenemethylsulfonylureas [64], pyrazolemethyl-sulfonylureas [65, 66], pyridylmethylsulfonylureas [60] and others [67, 68].

An alternative procedure that can be employed, which circumvents the need for an oxidation step, is the direct conversion of a benzyl halide to a benzylsulfonate (Fig. 40). The sulfonate can then be converted to a sulfonamide by treatment with phosphorous pentachloride or phosphorous oxychloride, to produce the sulfonyl chloride, followed by ammonia to yield the sulfonamide **110** [69]. The conversion of a benzyl halide to the sulfonate has be accomplished with sodium thiosulfate [69] or sodium sulfite [70]. The quenching of benzylic anions with sulfur dioxide or sulfuryl chloride has also been utilized as a method

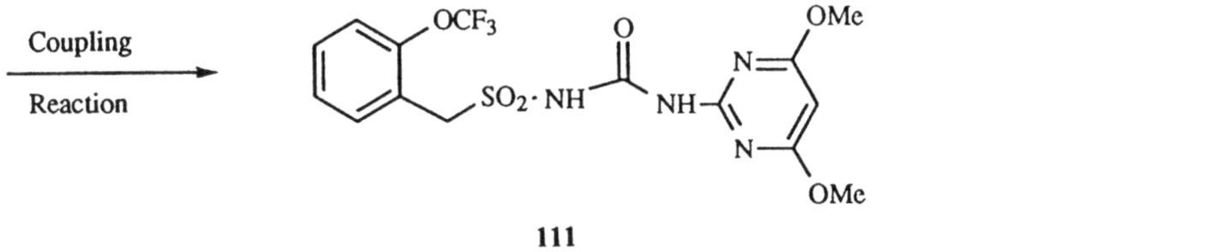

Fig. 40. Benzylsulfonylureas from benzylsulfonates

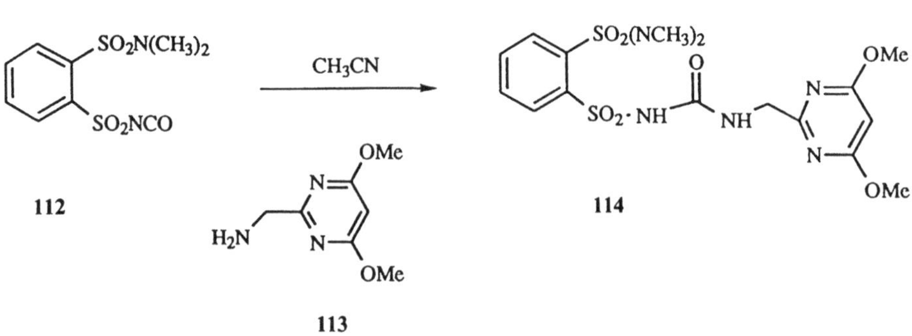

Fig. 41. 2-Aminomethylpyrimidinesulfonylureas

for the direct introduction of benzylsulfonates [23] and benzylsulfonyl halides [71].

Other than the original work reported by Levitt [72] (Fig. 41), bridge extended sulfonylureas in which a carbon has been inserted between the heterocycle and the sulfonylurea bridge (Fig. 29, Z = C) have not been examined further. The bridge extended sulfonylureas from the present case were obtained by reaction of 2-pyrimidinylmethylamine **113** with a sulfonylisocyanate. 2-Pyrimidinylmethylamines such as **113** are prepared via catalytic hydrogenation of the 2-cyanopyrimidine (Pd/C, Methanol/HCl).

3.2.2 Heteroatom Based Bridge Extensions (Fig. 29, Y or Z = O, N, S), Sulfamates (Y = 0), Sulfamoylureas (Y = N)

A considerable body of work has been published regarding the synthesis of sulfamates and sulfamoylureas. The methodology employed for the synthesis of these classes of compounds is however generally limited to two basic synthetic methods (Fig. 42). In the first method (Fig. 42 eq.1) a heterocyclic chlorosulfonylurea **116** is treated with either a phenol or aniline in the presence of a mild base to yield the sulfamates **117a** or sulfamoylureas **117b** respectively. The intermediate chlorosulfonylureas **116** are prepared from the heterocyclic amine by reaction with chlorosulfonylisocyanate. This methodology has been applied to the synthesis of arylsulfamates [73, 74] and a wide array of sulfamoylureas such

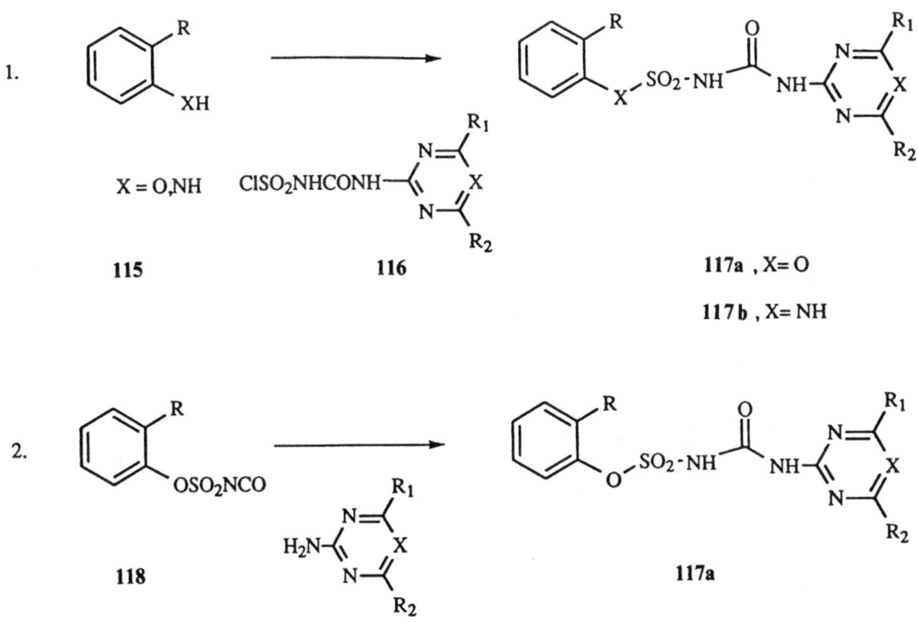

Fig. 42. Synthesis of sulfamates and sulfamoylureas

as simple arylsufamoylureas [74, 75], 2-hydroxyethylphenylsulfamoylureas [76], 2-cyclopropylcarbonylphenylsulfamoylureas [77] and pyrazolesulfamoylureas [78]. In the second method phenoxysulfonylisocyanate **114** combines with a heterocyclic amine to yield the expected sulfamates **117a**. Phenoxysulfonylisocyanates can be prepared, as reported by Lohaus [79], by the reaction of a phenol with chlorosulfonylisocyanate at elevated temperatures. A large portion of the sulfamates prepared to date have been synthesized utilizing the alkoxysulfonylisocyanate route. Examples include phenoxysulfamates [80, 81], 2,6-disubstituted-phenoxysulfamates [82], benzofurans [83] and haloalkoxysulfamates [84].

4 Synthesis of Bridge Modified Sulfonylureas (Sulfonylisoureas, Sulfonylisothioureas, Sulfonylguanidines)

While several reviews have detailed the synthesis and biological activity of "normal" sulfonylureas [23, 37], no reports have appeared which describe the synthesis of the less well represented structural variations such as the sulfonylisoureas, sulfonylisothioureas and sulfonylguanidines. This lack of coverage is undoubtedly due at least in part to the tremendous structural diversity allowed even within the context of the normal bridge sulfonylureas and the concentration of the synthetic effort around those areas with the highest levels of herbicidal activity. This section outlines the methods employed in the synthesis of several of the more prominent structural classes within the area of bridge modified sulfonylureas (Fig. 43).

4.1 General Methods for Sulfonylisoureas and Sulfonylisothioureas

The single most common method utilized in the preparation of sulfonyliso- and sulfonylisothioureas is the reaction of an *N*-sulfonyliminocarbonic ester **119** [85] with a heterocyclic amine. Typically the anion of the heterocyclic amine is

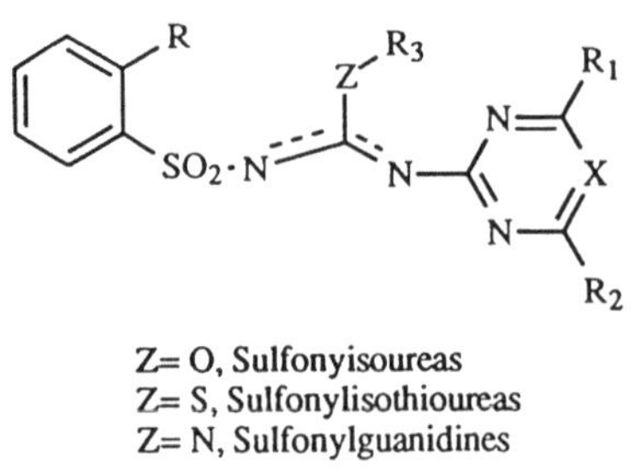

Fig. 43. Bridge modified sulfonylureas

required in these reactions. Sulfonylisoureas bearing an alkyl group on the bridge oxygen (Fig. 44, **120a**, R_3 = alkyl) have been prepared by the reaction of heterocyclic amines with *O*-alkyl *S*-alky sulfonyliminothiocarbonates (Fig. 44, Z = O, Y = S; R_3,R_4 = alkyl) [86]. Alternatively the sulfur of the *O*-alkyl *S*-alkyl sulfonyliminothiocarbonate can first be converted to a chlorine using sulfuryl chloride and the resultant *O*-alkyl sulfonyliminochloridate employed in the above coupling reaction [87].

Sulfonylisoureas bearing an aryl group on the bridge (Fig. 44, **120a**, R_3 = aryl) are prepared in a similar fashion utilizing diphenyl sulfonyliminocarbonates (Fig. 44, **119**; Z, Y = O; R_3, R_4 = aryl) [88]. The diphenyl sulfonyliminocarbonates used in this work were prepared from the corresponding dimethyl iminodithiocarbonates by first conversion to the sulfonyliminocarbonic acid dichlorides with sulfuryl chloride [88] or chlorine [88, 89] and then displacement of the chlorines with an appropriate phenol (sodium phenolate). Sulfonylisothioureas (Fig. 43, Z = S) can also be prepared utilizing this same general scheme.

Sulfonylisothioureas containing either an alkyl or aryl group on the isothiourea bridge (Fig. 44, **120b**, R_3 = alkyl or aryl) are readily prepared by reaction of sulfonyliminodithiocarbonates **119** (Z, Y = S; R_3, R_4 = alkyl or aryl) with a heterocyclic amine [90–94]. For the most part this synthetic scheme has only been applied to the synthesis of sulfonylisothioureas where the aryl portion of the molecule is a substituted phenyl ring. The sulfonyliminodithiocarbonates used in this work can be obtained (A) from the reaction of a sulfonamide with carbon disulfide followed by alkylation [91], (B) by displacement of chlorine from sulfonyliminocarbonic acid dichlorides with thiolates [94], and (C) by condensation of iminodithiocarbonates with sulfonyl halides [95].

Sulfonylisothioureas with an aryl group on sulfur (120b, R_3 = aryl) have similarly been synthesized from *S*-aryl *O*-aryl iminothiocarbonates by condensation with heterocyclic amines [96].

Another procedure that has been successfully employed for the synthesis of sulfonylisoureas and sulfonylisothioureas is the reaction of heterocyclic isoureas and isothioureas respectively with sulfonyl halides (Fig. 45) [97]. This method has made possible the synthesis of a wide range of structural variation in the aryl

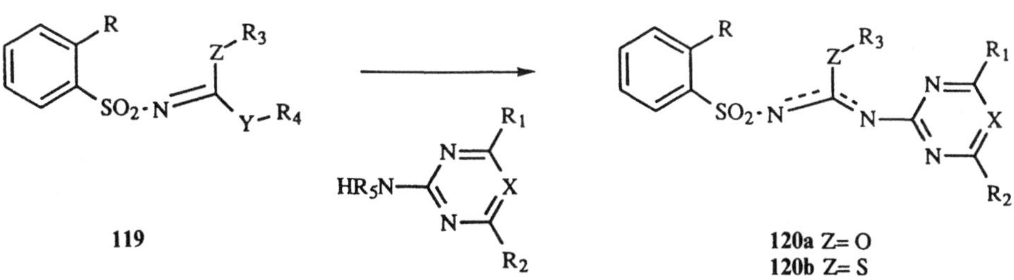

119

120a Z= O
120b Z= S

Fig. 44. Sulfonylisoureas and sulfonylisothioureas from *N*-sulfonyliminocarbonic ester derivatives

portion of both sulfonylisoureas and sulfonylisothioureas due to the ready availability of a large array of arylsulfonyl chlorides **121**. Sulfonylisoureas [98] and sulfonylisothioureas [97, 99] where the aryl group is phenyl, benzyl, pyrazole and thienyl have been prepared employing this methodology. The heterocyclic isoureas and isothioureas used in these studies are generally prepared by condensation of isourea or isothiourea with a halo substituted heterocycle. A detailed description of the synthesis of the heterocyclic isoureas and isothioureas can be found in the preceeding references [97–100].

Another method that has been applied to the synthesis of both sulfonylisoureas and sulfonylisothioureas is the displacement reaction of *N*-sulfonylguanidines with either alkoxides or thioalkoxides (Fig. 46) [101, 102]. The R-group (R_3) of the alkoxide or thioalkoxide may be alkyl, aryl or benzyl. This route has successfully been used to examine a wide range of aryl and substituted aryl groups on the bridge position of sulfonyliso(thio)ureas (R_3 = aryl).

Specifically substituted benzodisultams have also been exploited as substrates for this reaction leading to 2-sulfonamidophenylsulfonyliso(thio)ureas upon treatment with an alkoxide or thioalkoxide [103].

Sulfonylisoureas have also been prepared from sulfonylureas via first conversion to a carbaimidoyl chloride (via carbon tetrachloride/triphenylphosphine) and then displacement of the chloride with an alkoxide [104].

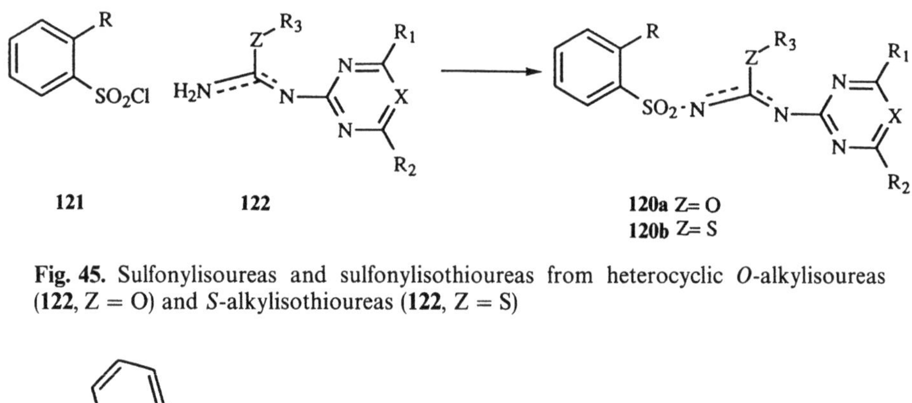

121 **122** **120a** Z= O **120b** Z= S

Fig. 45. Sulfonylisoureas and sulfonylisothioureas from heterocyclic *O*-alkylisoureas (**122**, Z = O) and *S*-alkylisothioureas (**122**, Z = S)

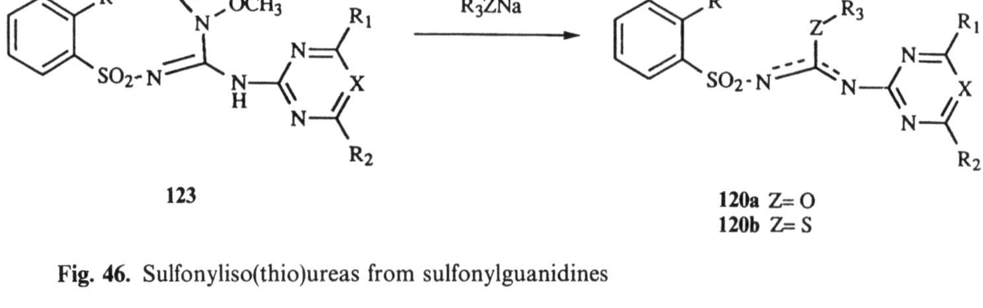

123 **120a** Z= O **120b** Z= S

Fig. 46. Sulfonyliso(thio)ureas from sulfonylguanidines

4.2 Specific Methods for Sulfonylisothioureas

The preceeding section outlined general methods used in the synthesis of both sulfonylisoureas and sulfonylisothioureas. Several additional methods have been reported which are specific for the synthesis of sulfonylisothioureas. Sulfonylthioureas have been converted to sulfonylisothioureas by alkylation [90]. Sulfonylthioureas when alkylated with chloroacetyl chloride form sulfonylisothioureas which are part of an oxothiazolidine ring system [105]. *N*-heteroaryl-iminodithiocarbonic acid *S,S*-diesters upon treatment with the salt of a sulfonamide provide the expected sulfonylisothioureas [106].

Sulfonylthioureas can also readily be converted to sulfonylisothiourea disulfides by treatment of the thioureas with alkylthioureas to give alkyldisulfides or aryl sulfenes/sulfilimines to provide the aryldisulfides [107]. Sulfonylthioureas can also be converted to the dimeric sulfonylisothioureas by treatment of the sulfonylthiourea salts with bromine [108].

4.3 Methods for Arylsulfonylguanidines (Fig. 47, R_4 = H, alkyl, aryl)

Probably the single most widely exploited method used in the preparation of simple sulfonylguanidines (Fig. 47, R_4 = H, alkyl) is treatment of sulfonylisothioureas with an amine. While this reaction is particularly well known for its widespread use in the synthesis of oximino ($R_4 = OR_5$) [97] and hydrazino derivatives ($R_4 = NR_6R_7$) [109] it has also found utility in the preparation of the simple alkyl sulfonylguanidines [97, 110–114]. Typically the sulfonylguanidines are prepared by heating in an inert solvent an equal molar mixture of the

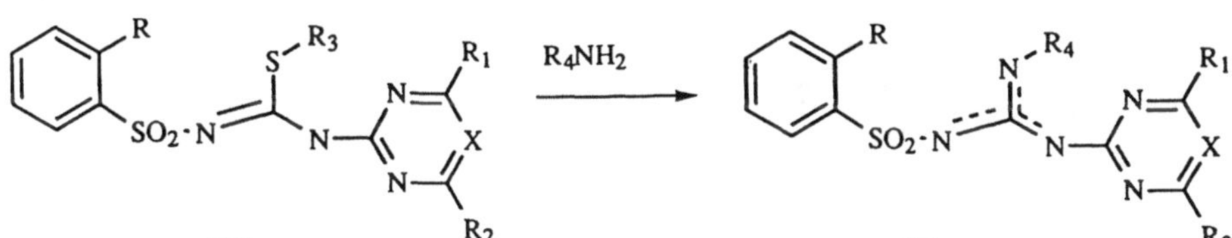

120b **124**

Fig. 47. Sulfonylguanidines from sulfonylisothioureas

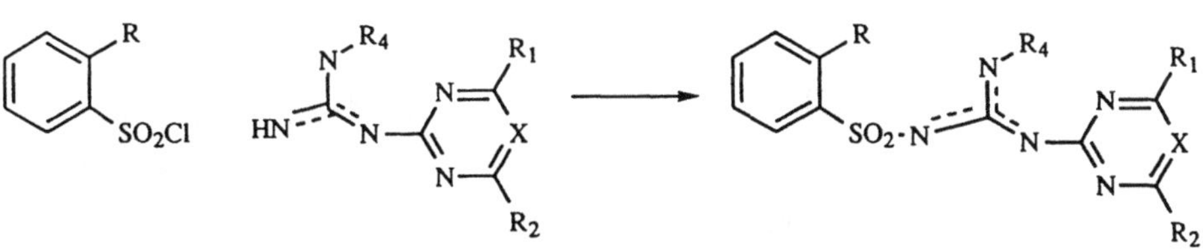

121 **125** **124**

Fig. 48. Sulfonylguanidines from heterocyclic guanidines and sulfonyl halides

sulfonylisothiourea with an appropriate amine. While this method has proven useful for the examination of modifications to the bridge substitutents (R_4), it does not lend itself to the examination of structural variations in the aryl portion of the molecule.

A method which can be used to explore a wide range of variation in the aryl portion of the molecule as well as substituents on the bridge (R_4) is the reaction of sulfonyl halides with heterocyclic guanidines (Fig. 48) [113–114]. The heterocyclic guanidines **125** used in this work are readily prepared by the addition of anilines, or simple alkyl amines to a cyanoaminoheterocycles [113–114].

Sulfonylguanidines have also been prepared from *bis*-benzenesulfonylguanidines by displacement with a variety of amines [113–114].

5 Synthesis of Bridge Fused Sulfonyliso(thio)ureas

Sulfonyliso(thio)ureas which have the bridge substituent linked back into the heterocycle forming a second ring (Fig. 49) have been investigated broadly by several groups [115–122]. While it may be interesting to speculate that these derivatives could in fact be pro-sulfonylureas, no evidence has published to date which addresses this question.

Essentially all of the sulfonylthiadiazoloazines **126a** prepared to date have been synthesized by oxidation of the corresponding sulfonylthioureas (Fig. 50)

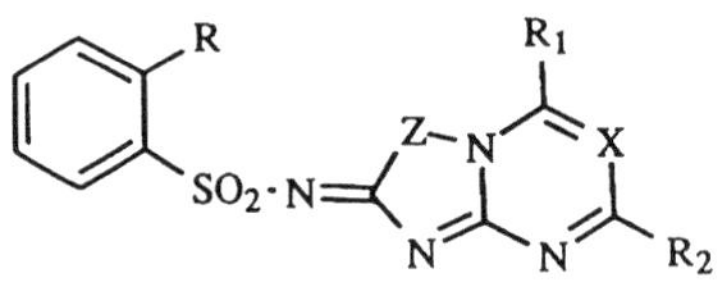

126a Z=S
126b Z=O

Fig. 49. Bridge fused sulfonyliso(thio)ureas

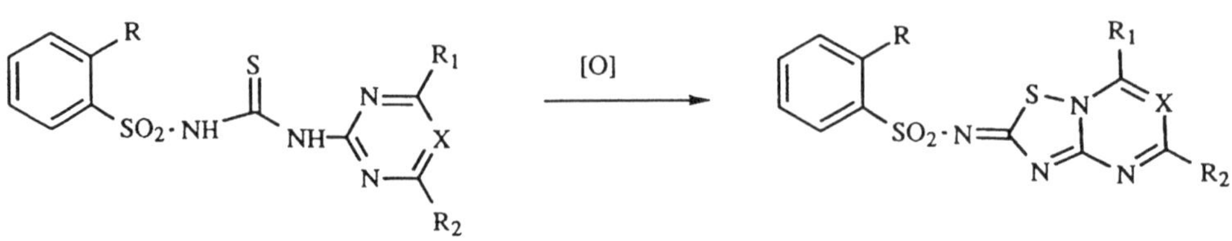

127 **126a**

Fig. 50. Sulfonylthiadiazoloazines by oxidation of sulfonylthioureas

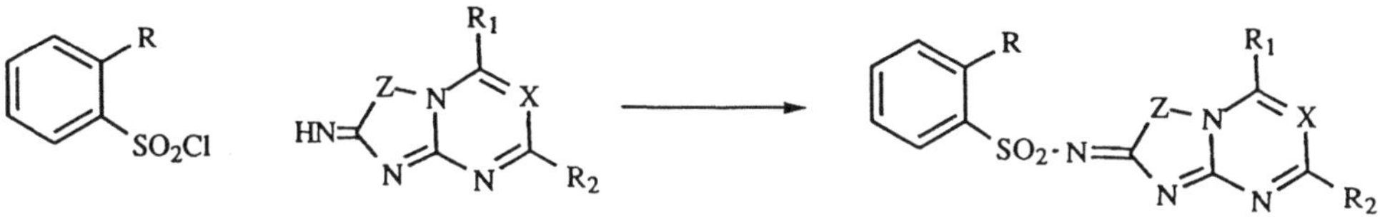

121 **128** **126a,b**

Fig. 51. Sulfonyloxa(thia)diazoloazines from condensation of sulfonyl halides and Oxa(thia)diazoloazines

[116–122]. These reactions are typically conducted by treating at low temperature a solution of the sulfonylthiourea with bromine [116], iodine [118], *N*-bromosuccinimide [122] or *N*-chlorosuccinimide. The cyclization proceeds very rapidly and typically in near quantitative yield. All functionality on the aryl group and the heterocycle appear to be tolerant to these reaction conditions.

A second method that is applicable to the synthesis of both the sulfonylthiadiazoloazines **126a** and sulfonyloxadiazoloazines **126b** [120] is the condensation of the oxa(thia)diazoloazines with arylsulfonyl halides [115] (Fig. 51). Treatment of an equal molar mixture of a sulfonyl chloride and an oxa(thia)diazoloazine **128** with triethylamine provides the sulfonyloxa(thia)diazoloazines **126a,b**.

Another method which is generally applicable to the synthesis of sulfonyloxadiazoloazines **126b** is the condensation of dimethyl sulfonyliminodithiocarbonates with heterocyclic 1-oxides [115].

6 Summary

Due to the vast amount of chemistry conducted in this area of herbicide research during the past twenty years, a comprehensive review of the chemistry of sulfonylureas is beyond the scope of this manuscript. We have however attempted to present an overview of the recent developments in the chemistry of sulfonylurea herbicides by highlighting the recent methods of synthesis of each integral part of the sulfonylurea molecule. Although research in sulfonylureas is diminishing across the crop protection chemical industry, some companies still seek to exploit this class of herbicides as evidenced by new patents continuing to publish. Undoubtedly new chemistry will continue to be developed as companies seek to discover sulfonylureas with new, more active *ortho*-substituents, heterocycles or bridge modifications.

7 References

1. Lombardino JG (1971) J Org Chem 36: 1843
2. Borrod G, Gradas A (1990) US 4960454
3. Hanagan MA (1987) US 4604131
4. Kimura F et al. (1986) JAP 152680/86
5. Kimura F et al. (1986) JAP 61-30585
6. Cuomo, J, Gee SK, Hartzell SL (1991) In: Baker DR, Fenyes JG, Moberg, WK (eds) Synthesis and Chemical of Agrochemicals II, American Chemical Society, Washington, pp 62–73
7. Levitt G (1984) US 4481029
8. Liang PH (1989) US 4808721
9. Lockett BL, Zimmerman WT (1992) WO 9204319
10. Tam WJ (1983) unpublished results
11. Hay JV, Wexler BA, Zimmerman DF (1987) US 4666501
12. Chen CW, Schwing GW (1983) US 4368069
13. Josey AD (1987) US 4683091
14. Hillemann CL (1986) US 4588432
15. Föry W (1992) WO 92216522
16. Kimura F (1989) US 4835282
17. Shapiro R (1988) US 4723988
18. Schuster R, Föry W, Meyer W (1985) US 4537618
19. Zimmerman WT (1987) US 4643761
20. Shapiro R (1986) US 4606754
21. Meyer W, Reinehr D, Oertle K, Schurter R (1987) EP 248245
22. Meyer W, Reinehr D, Oertle K, Schurter R (1988) US 4780125
23. Hay JV (1990) Pestic Sci 29: 247
24. Wolf AD (1986) US 4602940
25. Meyer W (1989) US 4857100
26. Christensen JR, Liang PH, Thompson ME (1987); Thompson ME, Liang PH (1991) In: Baker DR, Fenyes JG, Moberg WK (eds) Synthesis and Chemistry of Agrochemicals II, American Chemical Society, Washington, pp 87–97
27. Thompson ME (1987) US 4662931
28. Artz SP, Finkelstein BL, Hanagan MA, Moon MP, Pasteris RJ, Rorer MP, (1991) In: Baker DR, Fenyes JG, Moberg WK (eds) Synthesis and Chemistry of Agrochemicals II, American Chemical Society, Washinton, pp 50–61
29. Rorer MP (1989) US 4851030
30. Hartzell SL (1985) US 4705558
31. Wolf AD (1984) US 4465505
32. Rorer MP (1985) US 4511392
33. Kimura F et al. (1984) JAP 273152/84
34. Petersen WC (1987) US 4638000
35. Föry W (1986) US 4603211
36. Waldner A, Meyer W (1991) US 4999426
37. Levitt G (1991) In: Baker DR, Fenyes JG, Moberg WK (eds) Synthesis and Chemistry of Agrichemicals II, American Chemical Society, Washington, pp 16–31
38. Campopiano O, Moon M (1992) US 5157119
39. Kunz RA (1985) US 4546179
40. Georgescu E et al. (1988) RO 99497
41. Besenyei G et al. (1993) WO 93/10086
42. Besenyei G et al. (1990) US 4954628
43. Besenyei G et al. (1991) US 5068424

44. Haga T et al. (1989) US 4835282
45. Mayer H et al. (1991) US 5071470
46. Diehr HJ et al. (1987) US 4666505
47. Diehr HJ et al. (1987) US 4704158
48. Diehr HJ et al. (1989) US 4820337
49. Diehr HJ et al. (1985) ZA 85/6589
50. Diehr HJ et al. (1987) US 4659364
51. Kehne H (1993) US 5235050
52. Levitt G (1989) US 4818278
53. Budzinski JC (1983) US 4368067
54. Zimmerman DF (1988) US 4764610
55. Kristinsson H, Topfl W (1985) US 4521597
56. Shiokawa H et al. (1985) JAP 67469/85
57. Johnson TB, Sprague JM (1936) J Am Chem Soc 58: 1348; (1936) J Am Chem Soc 59: 1837 and 2439; (1939) 61: 176
58. Hass W et al. (1991) EP 447645
59. Johnson B, Douglas JB (1939) J Am Chem Soc 61: 2548
60. Liang P (1990) WO 90/00384
61. Sauers RF (1983) US 4420325
62. Hay JV, Levitt G (1987) US 4678500
63. Artz SP (1991) US 5071469
64. Kimura F et al. (1986) JAP 84/246725
65. Yamamoto S et al. (1984) JAP 85/214785
66. Wexler BA (1987) US 4693743
67. Pasteris RJ, Thompson ME (1986) US 4609395
68. Willms L et al. (1988) DE 3709340
69. Fest C et al. (1989) US 4806147
70. Reed RM, Tartar HV (1935) J Am Chem Soc 57: 571
71. Bhattacharya SN (1968) J Chem Soc C 1265
72. Levitt G (1985) US 4496392
73. Bohner B (1983) US 4391976
74. Tanaka Y et al. (1983) JAP 15962
75. Bohner B (1983) US 4515620
76. Van Gemert B (1987) US 4666508
77. Brady TE et al. (1991) US 5009699
78. Yamamoto S et al. (1985) JAP 214785
79. Lohaus G (1972) Chem Ber 105: 2791
80. Reap JJ (1980) US 4191553; (1983) US 4394153
81. Adams JB, Sauers RF (1986) US 4581059
82. Bauer K (1993) US 5272129
83. Rorer MP (1986) US 4576632
84. Willms L et al. (1984) US 4440565
85. Gompper R, Hagele W (1966) Chem Ber 99: 2885
86. Topfl W (1984) CH 646957
87. Schwing GW, Woods TS (1981) US 4301286
88. Fest C et al. (1989) US 4846879
89. Gompper R, Hagele W (1966) Chem Ber 99: 2900; (1965) Angew Chem 77: 430
90. Grantham GD, Levitt G (1981) CA 1107283
91. Topfl W et al. (1985) EP 151554
92. Fest C et al. (1990) US 4911754
93. Reibel H et al. (1990) US 4923505
94. Fest C (1988) EP 264019
95. Riebel H (1989) DE 3726424
96. Fest C (1990) DE 3841185

97. Shapiro R, Tseng CP (1984) EP 117014
98. Lantzsch R et al. (1992) US 5126454
99. Gesing E et al. (1991) EP 413221
100. Riebel H (1993) US 5250685
101. Diehr HJ et al. (1991) US 4988379; (1987) US 4647665
102. Fest C et al. (1990) US 4911752
103. Diehr HJ et al. (1987) US 4649364
104. Schwing GW, Woods TS (1981) US 4301286
105. Kuragano T et al. (1988) JAP 61-238791
106. Riebel HJ (1990) US 4968796
107. Kuragano T et al. (1988) JAP 61-238789
108. Kuragano T et al. (1988) JAP 61-238790
109. Fest C et al. (1993) EP 530616
110. Shiokawa et al. (1985) US 4559079
111. Saito J et al. (1986) US 4568381
112. Belai I et al. (1992) HU 59567
113. Moriya K et al. (1986) US 4602938
114. Moriya K et al. (1989) US 4844730
115. Tseng CP (1983) US 4411690
116. Aoki I et al. (1989) US 4872901
117. Okada Y et al. (1990) US 4897105
118. Hagiwara K et al. (1987) JAP 61-104532
119. Kimura F et al. (1989) JAP 62-287088
120. Kimura F et al. (1989) JAP 63-2080
121. Kimura F et al. (1989) JAP 63-65049
122. Bellina RF (1991) US 5069709

Recent Advances in Sulfonylurea Herbicides

H.M. Brown and J.C. Cotterman

DuPont Agricultural Products, Stine-Haskell Research Center,
E.I. du Pont de Nemours and Co., Inc., Newark, DE 19714, USA

The sulfonylurea herbicides were discovered in 1975 and immediately set a new standard for chemical weed control. The subsequent worldwide development effort has led to the commercialization of 18 different active ingredients in over a dozen major crops, and additional development candidates have been announced. Some of the reasons for the rapid commercial acceptance of sulfonylurea herbicides include their exceptionally low application rates (often a factor of 10–100 times lower than conventional herbicides), favorable environmental and toxicological properties, and compatibility with the trend towards postemergence weed control.

This chapter reviews recent developments in the biology and biochemistry of the sulfonylurea herbicides. The agricultural utilities of 12 new sulfonylureas commercialized or introduced since 1990 and significant new uses of existing commercial sulfonylureas are reviewed. These include tribenuron methyl, amidosulfuron, and NC330 in wheat and barley; metsulfuron methyl, cinosulfuron, pyrazosulfuron ethyl, and imazosulfuron in rice; rimsulfuron, NC319/MON12000, and CGA 152,005 in maize; and ethametsulfuron methyl, triflusulfuron, flazasulfuron, and rimsulfuron in oilseed rape, sugar beets, turf, and potatoes, respectively. Additionally, the biochemical crop selectivity mechanisms of five sulfonylureas in maize, wheat, and oilseed rape are reviewed. Metabolic reactions involved include aryl and aliphatic hydroxylation and oxidative O- and N-dealkylation leading to inactivation of herbicidal compounds or selective activation of proherbicidal sulfonylurea analogs. Finally, current information relating to weed resistance to sulfonylureas is reviewed including occurrence, mechanisms, fitness, gene flow, and management strategies. New developments include, among others, the occurrence of ALS-inhibitor resistant biotypes of aquatic species in U.S. rice production and the appearance of ALS-inhibitor resistant biotypes of *Xanthium strumarium* in soybeans. In addition, recent studies of competitiveness, seed production and other fitness comparisons suggest little or no difference in fitness between susceptible and resistant biotypes. Finally, management strategies are described which are designed to preserve the widespread utility of this important class of crop protection tools.

Chemistry of Plant Protection, Vol. 10

1 Introduction

The discovery of the sulfonylurea herbicides in 1975 signaled the beginning of a new era in chemical weed control technology and triggered a worldwide development effort that has led to the commercialization of 18 different active ingredients for selective weed control in over a dozen major crops. Some of the reasons for the rapid commercial acceptance of sulfonylurea herbicides include their low application rates, superior control of difficult weed species, good crop selectivity, environmental safety and compatibility with the trend towards postemergence weed control.

The chemistry, biology, and biochemistry of the sulfonylureas are the subject of a growing scientific literature, and these subjects have been occasionally reviewed in recent years [1–6]. One measure of the rapid commercial development of new sulfonylurea herbicides is that the most recent review [5] listed 15 sulfonylurea active ingredient products and candidates while just 3 years later one can cite 21 distinct active ingredients that are either commercialized or under development. In addition, understanding of the mode of action, basis for crop selectivity, environmental properties and development of weed resistance has grown sharply in recent years.

The objective of this chapter is to review recent developments in the biology and biochemistry of the sulfonylurea herbicides. New active ingredients commercialized or introduced since 1990 will be reviewed, as will significant new uses for existing actives. Recently published sulfonylurea crop selectivity mechanisms will be covered, as will developments related to sulfonylurea weed resistance.

2 New Sulfonylurea Herbicides: Agricultural Utility

The sulfonylurea herbicides were discovered in 1975 by Dr. George Levitt at DuPont [7]. Since that time, hundreds of patents covering tens of millions of herbicidal sulfonylureas have been issued to over two dozen agrichemical companies [5]. Perhaps more than any other herbicidal class, the sulfonylureas have yielded a broad diversity of products having distinct weed control spectra, crop selectivities, and suitabilities to different cropping systems. These utilities range from residual broadleaf weed control in dryland wheat production in North America, to broadleaf and sedge control in tropical rice paddies, to postemergence weed control in European sugar beets. The inherent flexibility of this chemical class is reflected in Tables 1–4 which show the structures and general agricultural utilities of 21 sulfonylurea active ingredients. The following sections describe the agricultural utilities of sulfonylureas which have been introduced since 1990.

2.1 Cereals

The widespread use of sulfonylurea herbicides for weed control in wheat and barley began with the commercialization of chlorsulfuron by DuPont in 1981. Since that time, five distinct sulfonylurea active ingredients have been commercialized for use in cereals, with a seventh currently under development (Table 1). These individual active ingredients address market needs that vary by region and agronomic practice. For example, chlorsulfuron [8–10], metsulfuron methyl [11–13], and triasulfuron [14–16] each offer residual control of broadleaf and specific grass weeds in wheat and barley while tribenuron methyl [17, 18] and thifensulfuron methyl [19, 20] are shorter residual products for this utility. These herbicides offer separate tools for cereals production systems ranging from dryland wheat/fallow rotations in North America and Australia to double cropped winter or spring cereals in high rainfall regions in Europe and the United States.

Three recent developments among cereal-selective sulfonylurea active ingredients are described below.

2.1.1 Tribenuron Methyl

Tribenuron methyl (formerly DPX-L5300) has been commercialized by Du Pont in N. America, Europe, Asia and other cereal producing countries under the tradenames "Express", "Granstar", "Pointer" and "Cameo". It is active postemergence against a broad spectrum of dicotyledonous weeds at 9–18 g a.i./ha. Important cereal weeds controlled by tribenuron methyl at this application rate include, among others, *Chenopodium album*, *Kochia scoparia*, *Cirsium arvense*, *Raphanus raphanistrum*, *Matricaria inodora*, *Sinapis arvensis*, *Stellaria media*, *Veronica persica*, *Viola arvensis*, and *Papaver rhoeas*. Tribenuron methyl is less effective against *Galium aparine*, *Veronica arvensis*, and *V. hederaefolia* and does not control monocotyledonous weeds in cereals [17].

Wheat and barley tolerance is excellent to application rates 2-fold greater than the highest labeled rates [17] with the exception of durum spring wheats and varieties of Mexican parentage. These varieties can exhibit moderate (10–20%) yellowing and/or stunting at application rates of 40 g a.i./ha [17]. This chlorosis and stunting can be overcome by tank mixture with labeled rates of phenoxy herbicides.

As seen in Table 1, tribenuron methyl is a close analog of metsulfuron methyl with a key difference being the *N*-methyl substituent on the sulfonylurea bridge of tribenuron methyl. This structural modification makes tribenuron methyl much more susceptible to abiotic hydrolysis of the sulfonylurea bridge [1, 3, 4]. For example, at pH values between 4–8, tribenuron methyl hydrolyzes 20–200 times faster than sulfonylureas having a nonmethylated bridge. This increased susceptibility to bridge hydrolysis results in very rapid degradation of tribenuron methyl in agricultural soils relative to other residual cereal sulfonyl-

Table 1. Cereal sulfonylurea herbicides

Chemical Structure	Common Name	Agricultural Utility	Application Rate (g ai/ha)
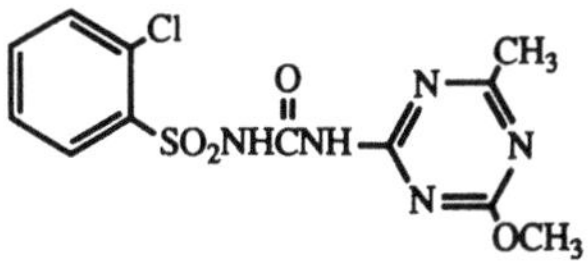	Chlorsulfuron (DuPont)	Broadleaf and select grass weeds	9 – 25 (post/pre)
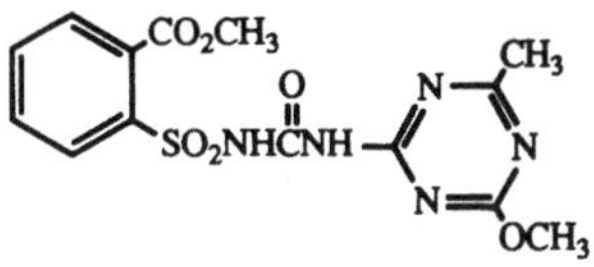	Metsulfuron methyl (DuPont)	Broadleaf weeds	4 – 7.5 (post)
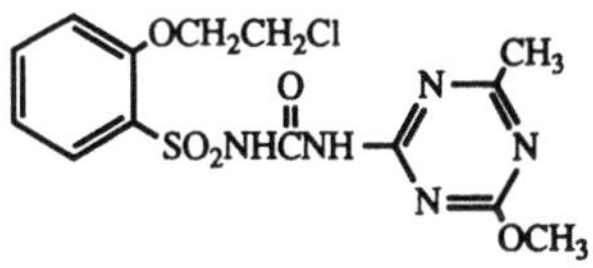	Triasulfuron (Ciba-Geigy)	Broadleaf and specific grass weeds	10 – 30 (post/pre)
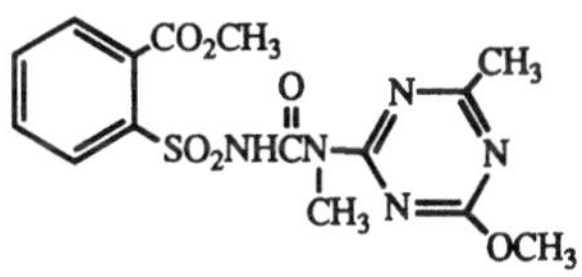	Tribenuron methyl (DuPont)	Broadleaf weeds	9 – 18 (post)
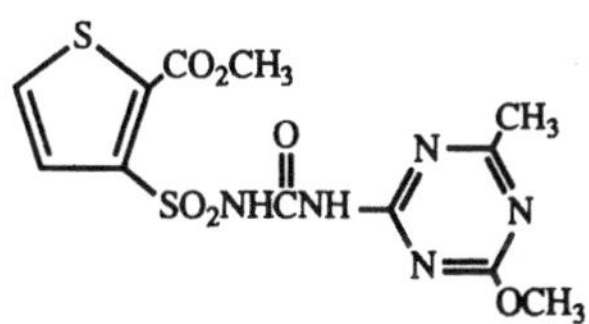	Thifensulfuron methyl (DuPont)	Broadleaf weeds	15 – 20 (post)
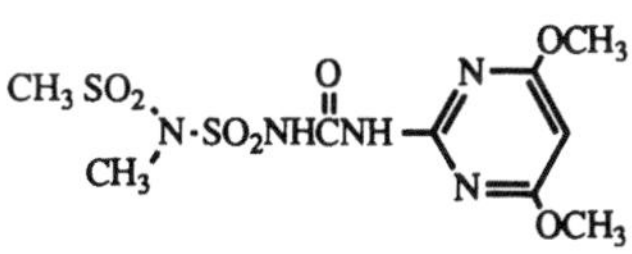	Amidosulfuron (Hoescht)	Broadleaf weeds (esp. Galium aparine)	30 – 60 (post)
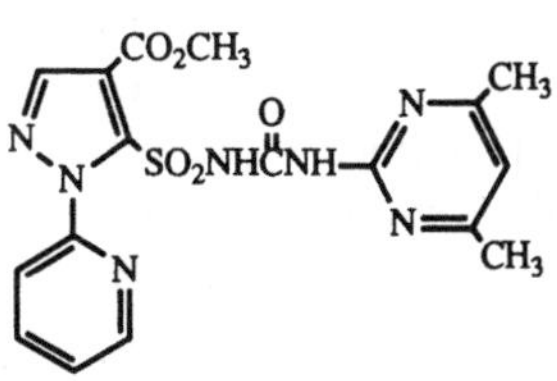	NC-330† (Nissan)	Broadleaf and select grass weeds	100 – 150 (post)

† Development Candidate

urea herbicides like chlorsulfuron. Thus, the short residual nature of tribenuron methyl is a key feature of this herbicide and offers utility in cereals production systems which include short recropping intervals to sensitive rotational crops.

2.1.2 Amidosulfuron

Amidosulfuron (formerly HOE 075032, Table 1) was commercialized in France in 1991 by Hoechst AG under the tradenames "Gratil" and "Adret" for postemergence broadleaf weed control in wheat and barley [21]. It is active against a select spectrum of broadleaf weeds at 30–60 g a.i./ha, with good control of *Polygonum* spp., *Raphanus raphanistrum*, and *Sinapis arvensis* at 30 g a.i./ha. It is generally less active against *Chenopodium album* and *Viola arvensis* at this rate [21]. Amidosulfuron offers special activity against *Galium aparine* at rates as low as 15 g a.i./ha. This feature distinguishes it from some other sulfonylureas in this market, where *Galium aparine* is often only suppressed by these products. Studies indicate good tolerance of spring and winter wheat and barley, rye, triticale, oats, and durum wheat to postemergence application rates of 90–120 g a.i./ha [21]. The mechanism of crop tolerance has not been published.

Amidosulfuron is atypical of sulfonylurea herbicides in that it does not contain a substituted aromatic sulfonamide. However, mode of action studies indicate that despite this marked structural deviation, amidosulfuron acts through inhibition of acetolactate synthase [22] as do all other sulfonylurea herbicides.

Degradation of amidosulfuron in soil is strongly dependent on temperature and moisture levels, and first-order half-lives in controlled laboratory experiments ranged from 26 ± 3 days to 79 ± 2 days in three Canadian soils at 20 °C [23]. Limited rotational crop sensitivity data is publicly available. Hoechst AG has commercialized amidosulfuron with some regionally-based rotational crop restrictions and it is not labeled for application with other sulfonylurea herbicides [21].

2.1.3 NC-330

NC-330 (Table 1) is a substituted pyrazole sulfonylurea introduced by Nissan Chemical Company for postemergence grass and broadleaf weed control in wheat. Two years of European field data show that postemergence application rates of 100–150 g a.i./ha of NC-330 gave good-to-excellent control of *Galium aparine*, *Stellaria media*, *Matricaria matricarioides*, and *Alopecurus myosuroides* with suppression of *Avena fatua* [24]. The activity against *A. myosuroides* is unique for cereal sulfonylureas. Field tests in the United Kingdom have confirmed this activity, where NC-330 at 100–200 g a.i./ha outperformed diclofop-methyl at 1134 g a.i./ha at two locations heavily infested with *A. myosuroides*

[24]. Crop tolerance work showed good selectivity on 12 cultivars of winter wheat at rates as high as 125 g a.i./ha, with better tolerance to applications at tillering than at early or late jointing stages. However, all seven cultivars of winter barley tested were moderately-to-severely injured by efficacious rates of NC-330 [24]. Available data indicate very low preemergence activity of NC-330. Detailed commercialization plans for NC-330 have not been announced.

2.2 Rice

Bensulfuron methyl was the first sulfonylurea herbicide commercialized for use in rice, and it remains the dominant sulfonylurea for broadleaf and sedge weed control in this crop [25, 26]. Table 2 includes the structures and utilities of four rice-specific sulfonylurea active ingredient products and candidates plus metsulfuron methyl, a cereal herbicide also labeled for use in rice in some countries.

2.2.1 Metsulfuron Methyl

Metsulfuron methyl has been commercialized in Europe and North America for postemergence broadleaf weed control in cereals (see Table 1) [11]. However, this sulfonylurea herbicide is also selective to transplant and direct seeded rice, especially *Indica* varieties, and it has been commercialized in Southeast Asia and Latin America for this utility [27, 28]. Application of 3–8 g a.i./ha metsulfuron methyl in paddy rice controls a broad spectrum of annual and perennial broadleaf weeds including *Monochoria vaginalis, Sphenoclea zeylandica, Selvinia molesta*, and *Marsilea crenata*. These application rates also provide some suppression of sedge weeds such as *Cyperus iria* but insufficient control of *Echinochloa* spp. Field studies show that crop tolerance to transplanted rice in Southeast Asia is excellent at rates between 8–16 g a.i./ha, with no effect on tillering, plant height or root length. However, application to direct-seeded rice prior to 5–7 days after seeding can lead to injury to the newly emerged seedlings [27]. The basis for rice selectivity to bensulfuron methyl is rapid metabolism [29], and it is likely that metabolism also accounts for rice selectivity to metsulfuron methyl.

Premixture with bensulfuron methyl (bensulfuron methyl: metsulfuron methyl, 5:1, w/w) provides increased activity against sedge weeds, and coapplication with 2,4-D also broadens the weed control spectrum [28].

2.2.2 Cinosulfuron

Cinosulfuron (formerly CGA-142464, Table 2) has been commercialized in Southeast Asia and other tropical regions by Ciba-Geigy for broadleaf and sedge weed control in paddy rice [30–33]. Greenhouse and field experience

Table 2. Rice sulfonylurea herbicides

Chemical Structure	Common Name	Agricultural Utility	Application Rate (g ai/ha)
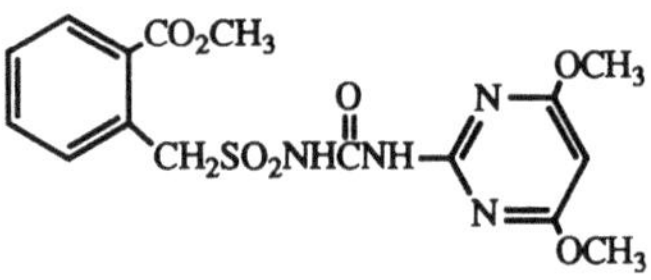	Bensulfuron methyl (DuPont)	Annual and perennial broadleaf and sedge weeds	20 – 75
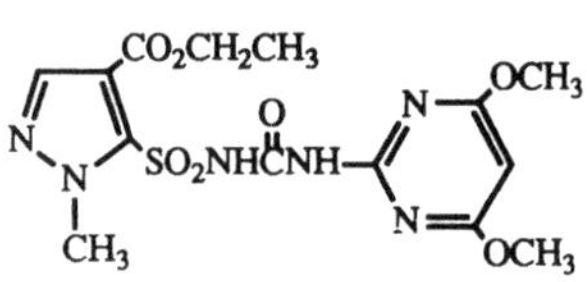	Pyrazosulfuron ethyl (Nissan)	Annual and perennial broadleaf and sedge weeds	20
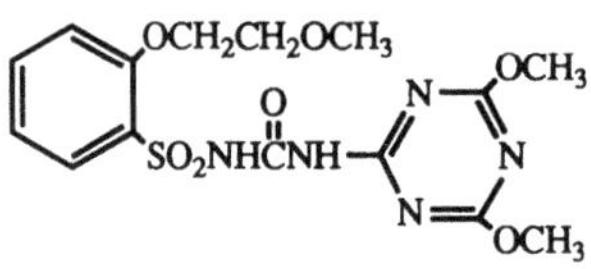	Cinosulfuron (Ciba-Geigy)	Specific annual and perennial broadleaf and sedge weeds	20 – 40
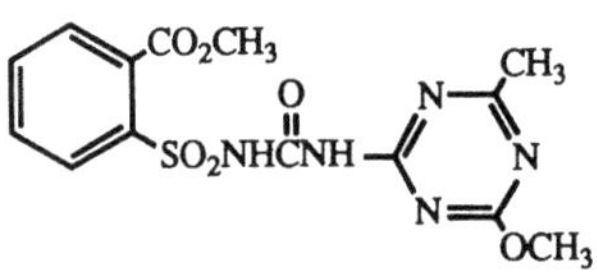	Metsulfuron methyl (DuPont)	Annual and perennial broadleaf and annual sedge weeds	2 – 6
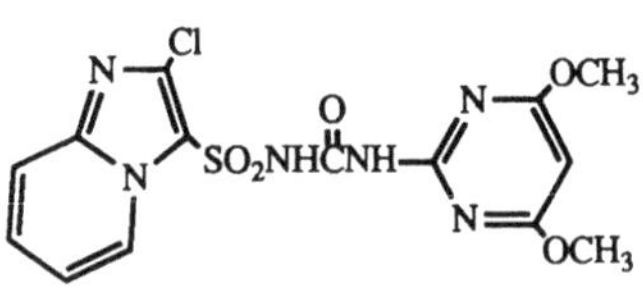	Imazosulfuron† TH-913 (Takada)	Annual and specific perennial broadleaf and sedge weeds	100

† Development Candidate

show that application at early timings of 20 g a.i./ha provide very good control of *Monochoria vaginalis, Marsilea crenata, Alisma* spp., select *Cyperus* and *Sagittaria* spp., and others. Later application timings require higher application rates up to 40 g a.i./ha. Moderately susceptible weeds include *Cyperus serotinus, Eleocharis kuroguwai, Fimbristylis* and *Ludwigia* spp., and only partial control of *Echinochloa* spp. Crop selectivity in permanently flooded and rain-fed paddies is dependent on application timing. Early injury (6–11% growth reduction) can occur at application rates of 10–40 g a.i./ha 3–9 days after transplant, although the treated crop recovers by 45 days after application [31]. Burhan et al. [32] reported that crop tolerance to 20 g a.i./ha cinosulfuron (applied as Setoff WG 20) was better than an alternative commercial treatment with 2,4-D EC 720.

2.2.3 Pyrazosulfuron Ethyl

Pyrazosulfuron (formerly NC 311, Table 2) was commercialized in 1990 by the Nissan Chemical Company for broadleaf and sedge weed control in paddy rice [34, 35]. Research shows that application rates ranging from 14–30 g a.i./ha can control a broad spectrum of annual and perennial broadleaf and sedge weeds including *Cyperus-*, *Alisma-*, *Sagittaria-*, and *Scirpus* spp. [36]. Application timing ranges from 3 to 20 days after transplanting. Pyrazosulfuron ethyl is active against *Echinochloa colona*, a barnyardgrass species relatively common in Southern Asia. However, it is less active against *E. crus-galli* and is sold as a premixture with other barnyardgrass herbicides to provide complete control of this and other *Echinochloa* species in countries such as Japan where the market demands a high degree of weed control [37]. Extensive field trials in Japan indicate good crop selectivity at 21 g a.i./ha when applied as 0.07% granules in transplanted rice. Higher rates required for barnyardgrass control provide a narrower margin of crop safety [37].

Pyrazosulfuron ethyl is sold under the trade name of Sirius as a 0.07% granule formulation and in premixtures with six barnyardgrass herbicides [38].

2.2.4 Imazosulfuron

Imazosulfuron (formerly TH-913, Table 2) was commercialized by Takeda Chemical Industries Ltd in 1993 under the tradename "Takeoff". Associated patent activity dates from 1987 covering imazosulfuron and related imidazo[1,2-*a*]pyridine and imidazo[2,1-*b*]thiazole sulfonylureas [39, 40]. Recent publications of outdoor concrete pot experiments [41, 42] show 93–100% control of broadleaf and sedge species including *Cyperus serotinus, Sagittaria pygmaea, Eleocharis kuroguwai, E. acicularis, Monochoria vaginalis* and *Cyperus difformis* at application rates of 50–100 g a.i./ha. Imazosulfuron is less active against *Echinochloa oryzicola* at these application rates. Rice safety was judged to be good, with 5–8% injury under the conditions of these tests. These biological properties suggest that, along with bensulfuron methyl and pyrazosulfuron ethyl, imazosulfuron will offer an additional tool for broadleaf and sedge weed control in the "one shot" Japanese rice herbicide market. Premixtures with other herbicides are under development, and planned tradenames are "Award", "Batl", "Hayate" and "Gosign".

2.3 Maize

A relatively recent development in sulfonylurea herbicides has been the discovery and development of maize-selective analogs. Although eight distinct cereal- and rice-selective sulfonylureas were commercialized during the 1980s, maize-selective sulfonylureas were not commercialized until 1990, when nicosulfuron and primisulfuron were introduced for postemergence grass and select broadleaf weed control in this crop. Since that time, three new sulfonylurea actives have been introduced into this segment.

Table 3 includes the structures and utilities of five maize-specific sulfonylurea active ingredients plus the utility of thifensulfuron methyl, a cereal herbicide also labeled for use in this crop. The following section reviews the properties of three recently introduced maize-selective sulfonylureas.

2.3.1 Rimsulfuron

Rimsulfuron (formerly DPX-E9636, Table 3) was commercialized in Europe by DuPont in 1992 for postemergence grass and broadleaf weed control in maize [43, 44]. Labeled application rates range from single applications of 7.5 g a.i./ha to split applications totaling 20 g a.i./ha per season. When combined with an approved non-ionic surfactant, a wide spectrum of grass weeds are controlled, including *Sorghum bicolor*, *Sorghum halepense*, *Setaria*- and *Panicum* spp., *Echinochloa crus-galli* (up to 1 tiller), *Elytrigia repens*, *Lolium multiflorum*, and *Digitaria sanguinalis* (up to 3 leaves). Dicotyledonous weeds up to the 2–4 leaf growth stage are also susceptible, including *Amaranthus*-, *Chenopodium*-, and *Matricaria* spp., *Abutilon theophrasti*, *Galium aparine*, *Stellaria media*, and *Helianthus annuus*.

Maize selectivity is based on rapid metabolic inactivation, and a wide crop safety margin is evident at labeled rates when applied at the 1–6 leaf stage [43, 44].

Rimsulfuron degrades rapidly in water and soil through a novel chemical mechanism. For example, half-lives in water at 25 °C range from 4.6 days to 0.3 days between pH 5 and pH 9, respectively, and soil half-lives in laboratory studies range between 1.7–4.3 days in a variety of soils at 30 °C [43]. The most

Table 3. Maize sulfonylurea herbicides

Chemical Structure	Common Name	Agricultural Utility	Application Rate (g ai/ha)
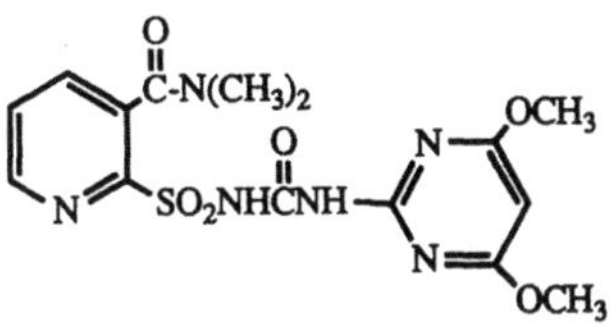	Nicosulfuron (DuPont and Ishihara)	Grass and specific broadleaf weeds	35 – 70 (post)
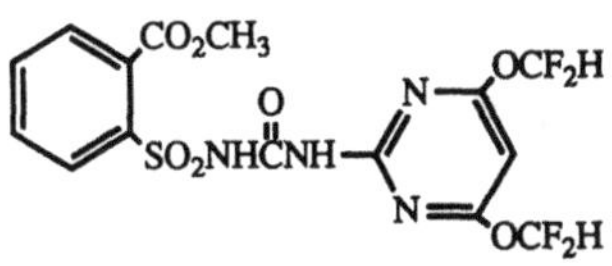	Primisulfuron (Ciba-Geigy)	Grass and specific broadleaf weeds	20 – 40 (post)
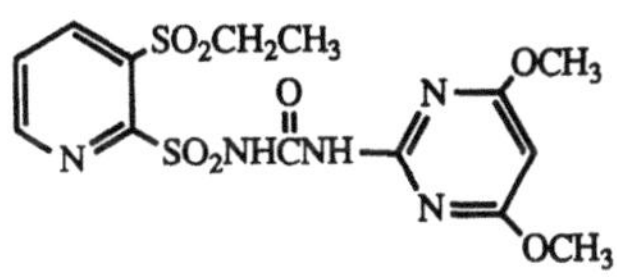	Rimsulfuron (DuPont)	Grass and broadleaf weeds	5 – 15 (post)
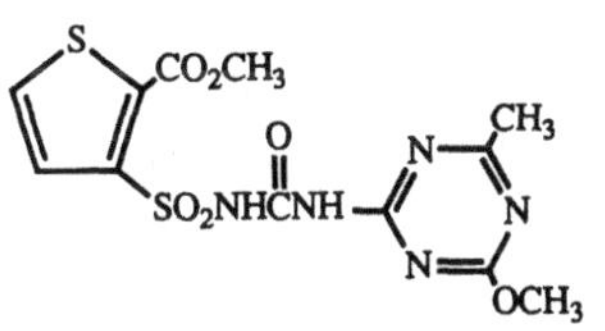	Thifensulfuron methyl (DuPont)	Broadleaf weeds	5 – 10 (post)
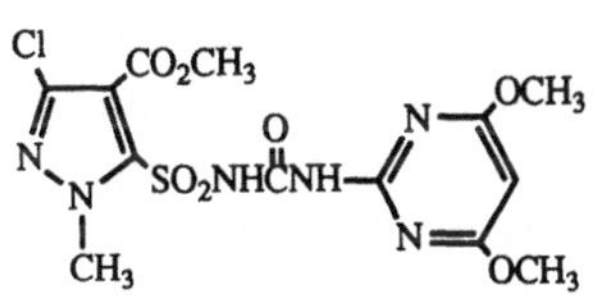	Halosulfuron† NC319/MON12000 (Nissan/Monsanto)	Broadleaf and sedge weeds	18 – 35 (post) 70 – 90 (pre)
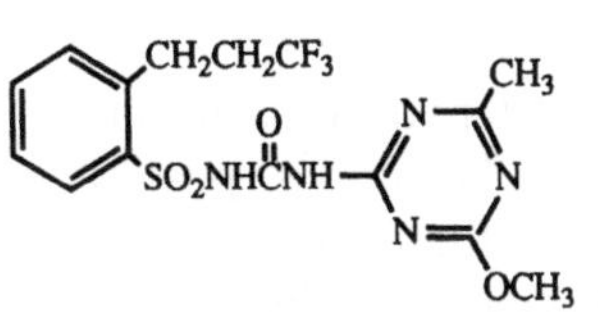	CGA-152005† (Ciba-Geigy)	Broadleaf weeds	20 – 40 (post) 15 (pre)

† Development Candidate

sensitive rotational crops likely to be planted after rimsulfuron applications in corn are lentils, sunflowers, and sorghum. Field trials with these and other rotational crops confirm that rapid degradation of rimsulfuron coupled with its low application rates allows for good rotational crop flexibility.

2.3.2 NC-319/MON 12000

NC-319 (Table 3) was discovered by Nissan Chemical Industries and is being co-developed with Monsanto Agricultural Products Company as MON 12000 for pre- and postemergence broadleaf weed control in maize. Greenhouse and field studies of postemergence applications of 20–40 g a.i./ha demonstrated good control of young seedlings of *Xanthium strumarium*, *Amaranthus retroflexus*, and *Abutilon theophrasti*, with higher rates required for late postemergence control of *A. theophrasti* [45–48]. MON 12000 is not active pre- or postemergence against graminacaeous weeds. Control of *Chenopodium album* was better post-emergence than preemergence, and maize tolerance was very good, with 0–10% injury reported at postemergence rates of 100 g a.i./ha on a number of dent corn cultivars.

In contrast, MON 12000 is less safe to maize preemergence or pre-plant incorporated, and plans call for co-development with the safener MON 13900 (furilazole) in a 1:3 herbicide:safener (wt:wt) ratio [49, 50]. Preemergence application rates of 63–70 g a.i./ha gave approximately 70–85% control of *X. strumarium*, *A. retroflexus*, *A. theophrasti*, and *C. album* when averaged over 2 years of US field trials, while inclusion of alachlor (3360 g a.i./ha) improved weed control to an average of 85–100%. *Ipomoea hederacea* and *I. purpurea* were not found to be effectively controlled by MON 12000.

Rotational cropping results to date show that in some trials sugar beet and sunflower were injured 12 months after preemergence treatments while soybeans and wheat were uninjured when planted 6–12 months after treatment with high application rates [50]. Thompson et al. [51] report that 3 years of field research indicate acceptable recropping safety to soybeans, spring or winter wheat, alfalfa, oats and sorghum when following 105–211 g a.i./ha MON 12000 plus furilazole safener. Sunflowers and sugar beets will require an 18–24 month rotational restriction [51]. Studies suggest that MON 12000 hydrolysis in water and soil is less affected by pH than other sulfonylurea herbicides, possibly because its very low pKa (approximately 1 unit lower than chlorsulfuron) insures that the large majority of soil residues are in the hydrolytically-resistant anionic form [51]. Additional studies showed that differential leaching and degradation of MON 12000 and its companion safener (furilazole) in soil may lead to spatial and temporal separation of these components in the soil profile under some conditions [52]. Further studies may be needed to determine whether such separation might affect the preemergence safening of this combination [52].

2.3.3 CGA-152005

CGA-152005 (proposed common name: prosulfuron) (Table 3) is under development by Ciba-Geigy for broadleaf weed control in maize [53–55]. Postemergence applications of 20 g a.i./ha were shown to give an average of > 85% control of *A. theophrasti*, *A. retroflexus*, *Ambrosia artemisiifolia*, *Polygonum convolvulus*, *X. strumarium*, and several other broadleaf weeds in maize. Species requiring 40 g a.i./ha for comparable control included *Ambrosia trifida*, *C. album*, *Ipomoea hederacea* and *I. lacunosa*. CGA-152005 is not active against grass or sedge weeds and is weak on *Solanum*- and *Sida* spp. Preemergence applications at 15 g a.i./ha are also under investigation for the control of certain small-seeded broadleaf weeds.

Results to date suggest very good postemergence tolerance of maize to CGA-152005, with metabolic inactivation being the likely basis for this tolerance. Maize is somewhat less tolerant to preemergence applications. For example, postemergence injury averaged from 1.8–2.4% at 40 g a.i./ha (113 observations) while preemergence application of this rate caused an average of 12% injury (143 observations).

Environmental fate and rotational cropping results to date suggest adequate degradation under field conditions, with faster degradation observed at increased soil temperature and moisture and decreased pH. Bareground application of 40 g a.i./ha in Hudson, New York showed CGA-152005 to undergo approximately 2–2.5 half-lives within the first 30 days after application (using a bioassay procedure), although the pH of the soil in this study was not specified [53]. As is common for sulfonylurea herbicides, there is a wide variation in tolerance to residues of CGA-152005 among rotational crops. Greenhouse studies showed that cereals, corn, sorghum, rice and field peas are tolerant to soil applications ranging from 12–100 g a.i./ha while sunflower, alfalfa and sugar beet are classified as sensitive at rates of 0.3–0.6 g a.i./ha (soil type not specified).

2.4 Other Crops

A majority of sulfonylurea herbicides have been commercialized for use in graminaceous crops. However, this broad class of chemistry has also yielded commercial products with selectivities towards broadleaf crops. Table 4 includes the structures and utilities of sulfonylurea herbicides with selectivities to soybeans, peanuts, oilseed rape, sugar beets and potatoes. Two others are also listed for non-crop weed control and use in managed turf.

2.4.1 Ethametsulfuron Methyl

Ethametsulfuron methyl (formerly DPX-A7881, Table 4) was commercialized by DuPont in 1989 for postemergence broadleaf weed control in Spring oilseed rape (*Brassica napus* L.) especially canola [56, 57]. Control of *Sinapis arvensis* L. (*Brassica kaber* [DC]) (wild mustard) and *Thlaspi arvense* (field pennycress) in canola is especially problematic and important since significant yield and oil quality reductions are caused by infestations by these weeds as low as 10 plants per square meter [58]. Field experience shows that application rates of 20–30 g a.i./ha ethametsulfuron methyl plus 0.2% non-ionic surfactant control *S. arvensis*, *T. arvense*, *Galeopsis tetrahit*, *Descurainia sophia*, and *Polygonum persicaria*, with suppression of *Amaranthus* spp. [56, 57, 59]. Crop tolerance to ethametsulfuron methyl is based on differential metabolic inactivation [60–62], and postemergence application rates up to 120 g a.i./ha were shown to be safe in Canadian field trials. Additional tests showed good selectivity to multiple commercial *Brassica napus* cultivars at labeled application rates [63, 64].

Ethametsulfuron methyl degrades in soil through pH-dependent chemical hydrolysis and microbial degradation. Its octanol-water partition coefficient is higher than for some other sulfonylurea herbicides, and this translates into greater soil sorption and reduced soil activity against both weeds and rotational crops. Field recropping studies in Canada have defined a minimum of a 10-month rotational interval to Spring and Durum wheat, barley, oats and flax and a 22 month interval to lentils, peas, faba beans, and oilseed rape. The interval to oilseed rape is based on occasional instances of rotational injury in field experiments and may result from the fact that oilseed rape roots metabolize ethametsulfuron methyl much more slowly than the foliage, meaning that rape tolerance is much greater for postemergence applications (especially from the 2 leaf growth stage to bolting) than under conditions where sufficient residues of ethametsulfuron methyl may remain in the soil, causing injury during germination and early seedling growth [61].

Weed control spectrum and recropping considerations have limited the use of ethametsulfuron methyl primarily to use in Canada [56, 65].

2.4.2 Triflusulfuron Methyl

Sugar beet (*Beta vulgaris*) is among the most sensitive plant species to many commercial sulfonylurea herbicides. Thus, it is a testament to the crop selectivity flexibility of the sulfonylurea chemical class that a sugar beet-tolerant sulfonylurea herbicide is now under development. Triflusulfuron methyl (formerly DPX-66037, Table 4) is being commercialized by DuPont for broadleaf weed control in sugar beet and fodder beet [66–69]. Field studies in Europe and N. America show that two to three sequential postemergence applications of 15 g a.i./ha triflusulfuron methyl spaced at 5 to 15 day intervals plus non-ionic surfactant or crop oil concentrate control a broad range of weeds including *Galium aparine*, most *Polygonum* spp. (except *P. convolvulus*), *Matricaria* spp., *Sinapis arvensis*, *Solanum nigrum*, and *Mercurialis annua*. Weeds suppressed by triflusulfuron methyl include *Chenopodium album*, *Stellaria media*, *P. aviculare*, and *Alopecurus myosuroides*. Crop selectivity is based on rapid metabolic inactivation by sugar

Table 4. Other sulfonylurea herbicides

Chemical Structure	Common Name	Crop	Agricultural Utility	Application Rate (g ai/ha)
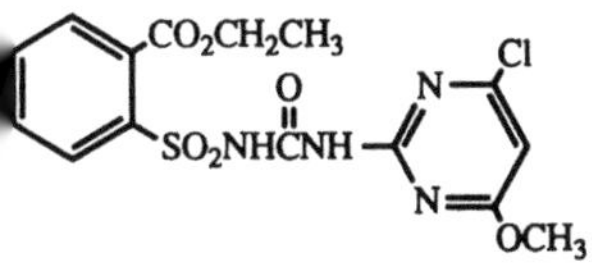	Chlorimuron ethyl (DuPont)	Soybeans Peanuts	Broadleaf weeds	8 – 13 (post) 35 – 70 (pre)
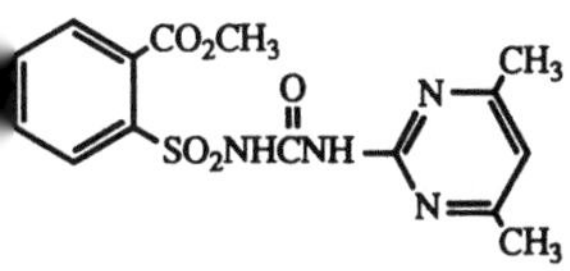	Sulfometuron methyl (DuPont)	Non-crop	Broadspectrum weed control	70 – 680 (post/pre)
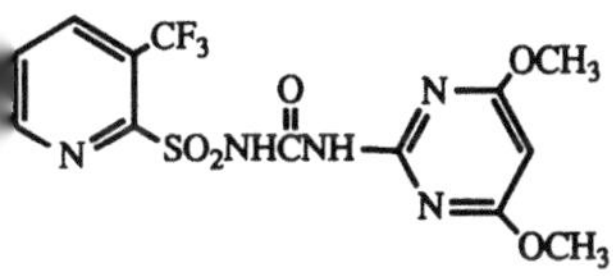	Flazasulfuron (Ishihara)	Managed Turf	Broadspectrum weed control	25 – 100 (post/pre)
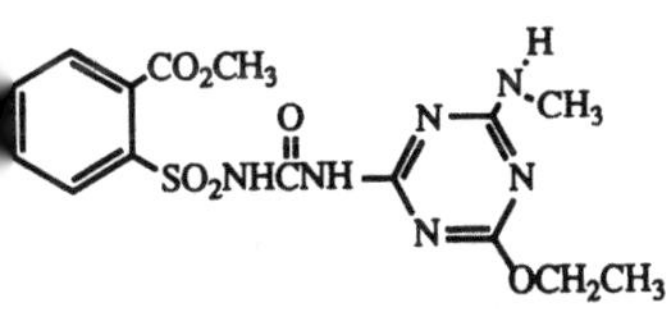	Ethametsulfuron methyl (DuPont)	Oilseed Rape	Broadleaf weeds	15 – 20 (post)
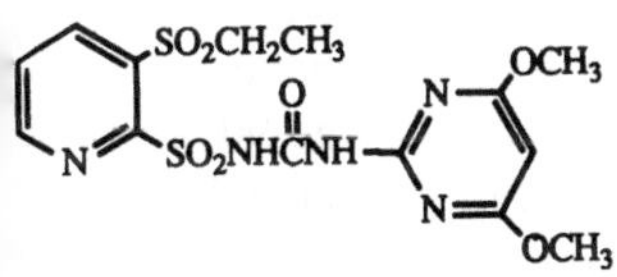	Rimsulfuron (DuPont)	Potatoes	Grass and broadleaf weeds	8 – 35 (post)
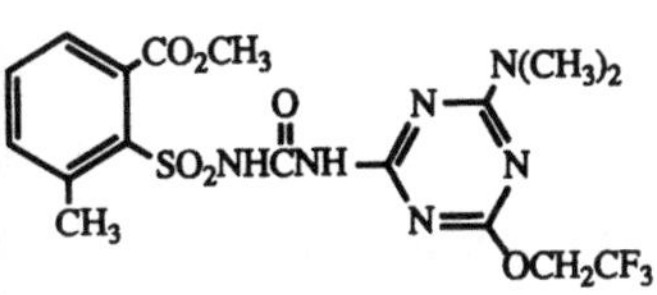	Triflusulfuron (DuPont)	Sugarbeets	Broadleaf weeds	15 (post)

beet foliage, and very good tolerance is reported at twice the expected use rate [66]. Hydrolysis half-lives of 3.7 days and 32 days at pH 5 and pH 7 (25 °C), respectively, contribute to rapid degradation in soil, and no rotational crop restrictions are anticipated. Triflusulfuron methyl will be commercialized in 1994.

2.4.3 Flazasulfuron

Flazasulfuron (formerly SL-160, Table 4) is being developed by Ishihara Sangyo Kaisha, Ltd. for broad spectrum weed control in managed turf grass such as on golf courses [70, 71]. It is active pre- and postemergence against annual grasses as well as annual and perennial broadleaf weeds and sedges. Early postemergence application timings are recommended, and rates of 25–100 g a.i./ha control many weeds including *Cyperus* spp., *Digitaria sanguinalis, Echinochloa crus-galli, Poa annua, Stellaria media, Polygonum lapathifolium,* and *Oxalis corniculata.* Soil residual activity against select annual weeds persists for several months in winter and about 30 days in summer.

Good tolerance is exhibited by certain turfgrasses including *Zoysia japonica, Z. matrella,* and *Cynodon dactylon,* and tolerance of these grasses to flazasulfuron is based on rapid metabolism. However, some warm-season and most cool-season turfgrasses are sensitive to flazasulfuron, and many are severely injured or killed at labeled application rates [71].

Tolerance of ornamental trees is a key consideration for this specialized herbicidal utility. Studies show that most evergreen trees show good-to-excellent tolerance up to 100 g a.i./ha applied to the foliage. An exception is *Viburnum awabuki.* Also, most deciduous trees are sensitive to flazasulfuron.

Flazasulfuron degrades rapidly in water and soil by a novel chemical mechanism [70]. For example, its half-life in water is 11 days at 25 °C and 90% of applied flazasulfuron degrades within 80–100 days at 15 °C in several soils.

Flazasulfuron is now commercialized as Shibagen in Japan.

2.4.4 Rimsulfuron (Potatoes)

The maize-selective sulfonylurea rimsulfuron (formerly DPX-E9636, Table 4) is also selective to solanaceous crops, especially potatoes [72–74] and is being commercialized for this utility by DuPont. Application rates of 7.5–15 g a.i./ha plus 0.125–0.25% non-ionic surfactant control annual and perennial grass weeds including *Digitaria sanguinalis, Echinochloa crus-galli, Elytrigia repens, Setaria* spp., *Panicum* spp., and others. Broad-leaved species that are controlled include *Galium aparine, Matricaria-, Sinapis-,* and *Amaranthus* spp., *Stellaria media,* and *Chenopodium album.* Established perennial grass weeds should have adequate growth and foliage at the time of treatment to intercept and absorb the active ingredient.

Selectivity in potatoes is based on differential metabolism, and potato cultivars have shown good tolerance at application rates up to 140 g a.i./ha. Newly emerged potatoes are more sensitive than established plants, and early treatments show good tolerance up to 25 g a.i./ha. Later treatments to established potatoes show excellent tolerance, and even at rates as high as 140 g a.i./ha, there were no effects on yield nor differences in tuber quality or seed potato vigor.

The rapid degradation of rimsulfuron in soil allows for full recropping flexibility. Rimsulfuron is currently commercialized for use in potatoes in several European countries with registration anticipated in Canada and the United States.

3 Plant Metabolism of Sulfonylurea Herbicides

Sulfonylurea herbicides have been commercialized or are in development for selective weed control in at least 12 major food or amenity crops including wheat, barley, oats, maize, rice (including both *japonica* and *indica* varieties), select turf grasses, soybeans, oilseed rape, flax, sugar beets, potatoes and tomatoes. In addition, each of these sulfonylurea herbicides has a distinct spectrum of weed control, including some weeds that are also tolerant to each respective herbicide. In each case where the mechanism of plant selectivity has been investigated, plant metabolic inactivation has been found to be the basis for the observed tolerance. The following section reviews recent developments in our understanding of the plant metabolism of the sulfonylurea herbicides and is intended to be additive to recent reviews on this subject [1, 4–6].

3.1 Nicosulfuron

As seen in Table 3, a number of new sulfonylureas have been discovered for postemergence weed control in maize. Nicosulfuron [75–77] was commercialized in the early 1990s for postemergence grass and select broadleaf weed control in maize. The metabolic basis for maize tolerance to nicosulfuron has been determined [60, 75]. Laboratory studies show that nicosulfuron is metabolized in maize with half-lives ranging from 1.5–4.5 hrs, whereas sensitive species including *Sorghum*- and *Setaria* spp. metabolize the herbicide much more slowly. Figure 1 summarizes the metabolic pathway of nicosulfuron in maize. Maize metabolizes nicosulfuron on the 5-position of the pyrimidine ring, and this metabolite is inactive against acetolactate synthase. This hydroxylated metabolite is subsequently rapidly conjugated to glucose. Structure/activity studies show that maize selectivity is strongly affected by substituents on the

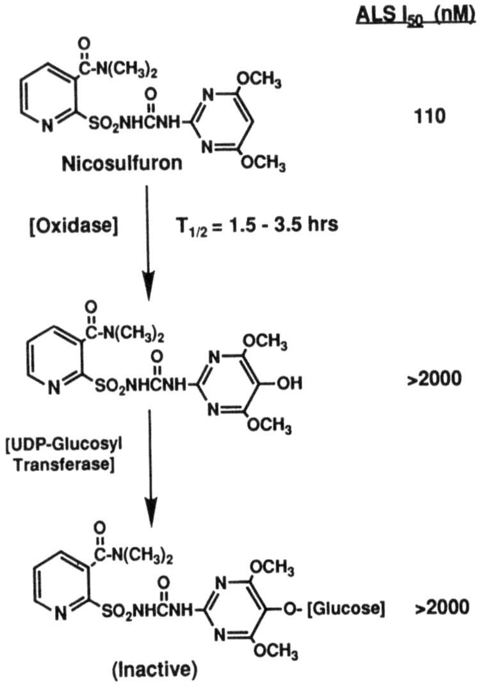

Fig. 1. Metabolic inactivation of nicosulfuron in maize. ALS I_{50} is the concentration inhibiting acetolactate synthase 50% in vitro

pyridine half of the molecule, a site relatively distant from the actual site of metabolism. Examples cited show that the *ortho*-carboxymethyl- and *ortho*-methylsulfonyl pyridine analogs of nicosulfuron are highly injurious to maize, and by inference, not effectively recognized as substrates by this or other xenobiotic metabolizing enzyme systems in maize.

3.2 Primisulfuron

Primisulfuron is another new sulfonylurea herbicide recently commercialized for postemergence grass and broadleaf weed control in maize [78, 79]. Figure 2 summarizes the metabolism of primisulfuron in maize [80] and *Echinochloa-crus galli* (L.), a tolerant weed [81]. Each of these tolerant species metabolizes primisulfuron rapidly, with half-lives of 1.5 h (*E. crus-galli*) and 5 h (*Zea mays*), and as for nicosulfuron, both species metabolize this sulfonylurea primarily via hydroxylation of the 5-position of the pyrimidine ring with subsequent conjugation to glucose. In addition, *E. crus-galli* (L.) also produces a non-glucosylated metabolite (not shown), and maize hydroxylates primisulfuron on the phenyl

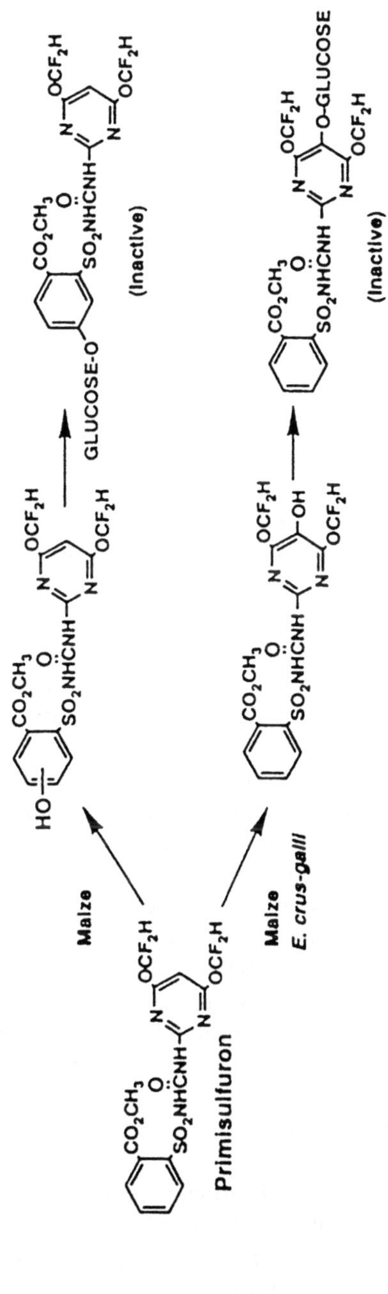

Fig. 2. Metabolism of primisulfuron in maize and *Echinochloa crus galli.* (L.)

ring (position not determined), and this metabolite is also conjugated to glucose (see Fig. 2). This latter reaction is analogous to that of chlorsulfuron and metsulfuron methyl in wheat [82, 83]. Based on work with other sulfonylureas (including nicosulfuron), both conjugated forms of primisulfuron are expected to be herbicidally-inactive [see 4]. In addition, Fonne-Pfister et al. [80] have shown that microsomal preparations from maize produce both hydroxylated metabolites shown in Figure 2 in vitro and that the characteristics of these enzymatic reactions are consistent with a cytochrome P450 monooxygenase system.

3.3 Triasulfuron

Six sulfonylurea herbicides have been commercialized for selective weed control in wheat and barley (Table 1), and the metabolic basis for cereal selectivity has been described for chlorsulfuron [82], metsulfuron methyl [83], thifensulfuron methyl [84], and more recently, for triasulfuron. Meyer and Muller [85] compared the uptake, translocation, and metabolism of ^{14}C-triasulfuron in tolerant wheat and sensitive *Lolium perenne.* There was little difference between these species in foliar uptake and translocation, and while more triasulfuron was taken up by wheat roots than roots of *L. perenne*, more radioactivity was translocated from the roots to the leaves in *L. perenne.* Based on these conflicting trends, these authors concluded that differential uptake and translocation cannot account for this crop selectivity. However, as for other cereal-selective sulfonylureas, wheat metabolized triasulfuron much more rapidly than this sensitive weed, with 14% of the applied triasulfuron remaining intact after 24 h in wheat compared to 70% in *L. perenne.* The primary metabolic pathway for triasulfuron in wheat was also described and is summarized in Fig. 3. Note that rapid phenyl hydroxylation and conjugation to glucose is also the basis for wheat selectivity to chlorsulfuron and metsulfuron methyl, but this reaction is not the basis for tolerance of this crop to thifensulfuron [84]. Frear et al. [86] have isolated microsomal fractions from wheat foliage that catalyze the hydroxylation of chlorsulfuron, triasulfuron and several other wheat-selective herbicides. Their studies suggest that wheat microsomes contain several different cytochrome P-450-linked monooxygenases having different properties and substrate specificities that are responsible for herbicide metabolism.

3.4 Ethametsulfuron Methyl

Ethametsulfuron methyl (Table 4) presents an interesting case of very distinct species selectivity. This sulfonylurea herbicide controls broadleaf weeds in spring oilseed rape (*Brassica napus*) including wild mustard (*Sinapis arvensis*, formerly *Brassica kaber*), a species that is so closely related to *B. napus* that these species can cross and produce viable progeny. Lichtner et al. [61] have

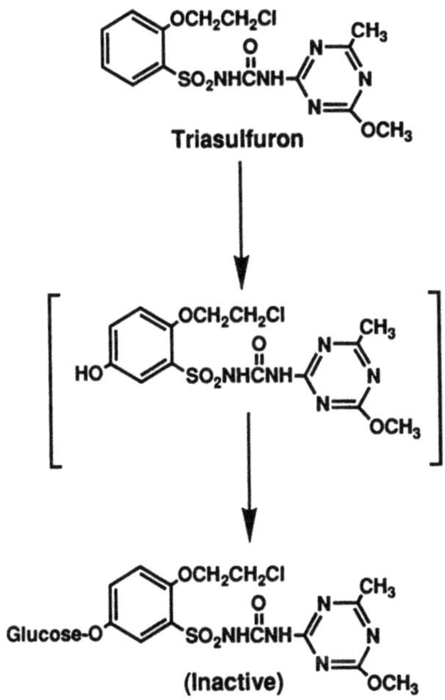

Fig. 3. Metabolic inactivation of triasulfuron in wheat foliage

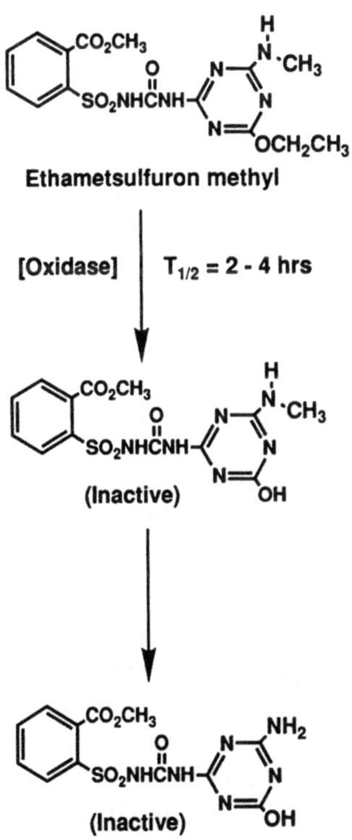

Fig. 4. Metabolic inactivation of ethametsulfuron methyl in foliage of oilseed rape (*Brassica napus* L.)

demonstrated a close correlation between plant sensitivity or tolerance to ethametsulfuron methyl and metabolism. Tolerant oilseed rape metabolizes this herbicide with a half-life of approximately 1–3 h and requires greenhouse application rates of 230 g a.i./ha to cause 50% injury. On the other hand, *S. arvensis* metabolizes ethametsulfuron methyl more slowly (half-life approximately 11 h) and exhibits 50% injury at rates as low as 16 g a.i./ha. Other observations include marginally faster metabolism and better postemergence tolerance in *B. napus* varieties versus *B. campestris* varieties and faster metabolism in oilseed rape foliage than in roots, correlating to better postemergence than preemergence tolerance.

B. napus metabolizes ethametsulfuron methyl through oxidative *O*-dealkylation of the 4-ethoxy substituent followed by *N*-demethylation of this metabolite (Fig. 4) [61]. The initial 4-hydroxy-6-*N*-methyl triazinyl metabolite is herbicidally-inactive and the hydroxy group is thought to tautomerize and does not undergo conjugation with glucose.

Brown et al. [60] have demonstrated that very close analogs of ethametsulfuron methyl are not selective to oilseed rape, implying a very strict substrate specificity for *O*-dealkylation in this plant. For example, neither the 4-*methoxy* analog of ethametsulfuron methyl nor the 4-ethoxy-6-methyl analog are selective to oilseed rape, indicating a distinct specificity for *O*-dealkylation of alkoxy groups at least two carbons in length and a "steering" role for the *N*-methyl substituent.

3.5 Activation of Proherbicidal Sulfonylureas

While metabolic inactivation by tolerant crops is the common mechanism for selectivity to sulfonylurea herbicides, several examples of metabolic activation of proherbicidal sulfonylureas by weed species have been described [6]. DPX-H6564 (Fig. 5) is a DuPont research lead that is selective to soybeans and active against grass and some broadleaf weeds at 16–70 g a.i./ha. Paradoxically, DPX-H6564 is inactive against acetolactate synthase, the established site of action of sulfonylurea herbicides, with an ALS $I_{50} > 2000$ nM. It was shown that sensitive plants including *Sorghum halepense* and *Zea mays* metabolize DPX-H6564 relatively rapidly, with half-lives of about 7–9 h, while soybeans did not measurably metabolize this compound, with greater than 95% remaining intact after 24 h. Figure 5 (A) shows that DPX-H6564 is metabolized by these sensitive plants to its 4-hydroxymethyl derivative, which is highly active against acetolactate synthase (ALS $I_{50} = 11$ nM). The high activity of this metabolite against ALS is consistent with other structure–activity information, where other substituents smaller than the methyl acetal lead to high inhibitory activity against ALS. Thus, the high herbicidal activity of DPX-H6564 results from its metabolic activation by sensitive plant species, and its tolerance to soybeans results from the inability of soybeans to similarly activate this proherbicide.

Another example of differential metabolic activation is found in DPX-L8747 (Fig. 5, B), another soybean-tolerant research lead from DuPont [6]. DPX-

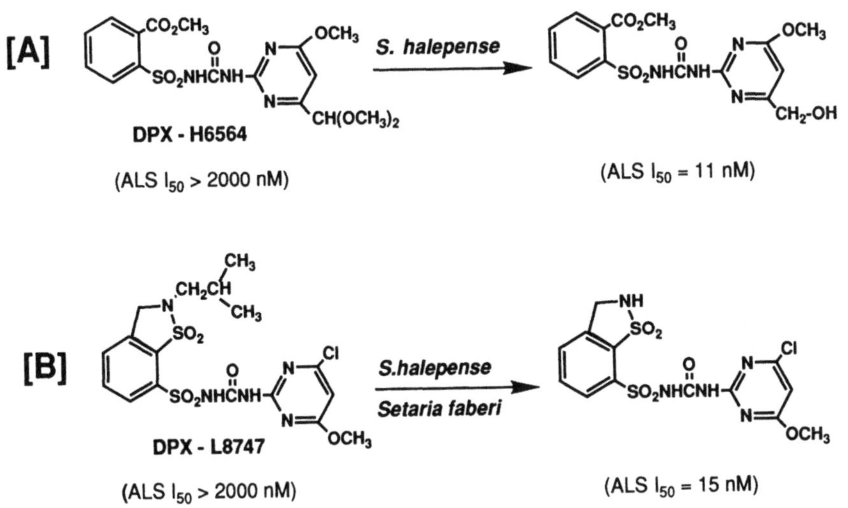

Fig. 5. Metabolic activation of proherbicidal sulfonylurea herbicides DPX-H6564 (A) and DPX-L8747 (B) in sensitive plant species

L8747 is inactive against ALS (ALS > 2000 nM) yet is highly herbicidal at 16–70 g a.i./ha. As for DPX-H6564, DPX-L8747 is metabolized by sensitive weeds including *Setaria faberi* and *Sorghum halepense* with half-lives of about 1–2 h. In this case, sensitive grasses *N*-dealkylate DPX-L8747 to form the hydrogen-substituted 1,2-benzisothiazole derivative (Fig. 5, B). This metabolite is highly active against ALS (ALS $I_{50} = 15$ nM) as are the *N*-methyl and *N*-ethyl analogs [6]. Tolerant soybeans also metabolize this proherbicide relatively rapidly ($T_{1/2}$ approximately 9–11 h). However, soybeans do not *N*-dealkylate DPX-L8747. Instead, soybeans catalyze its conjugation to homoglutathione at the 4-chloro position of the pyrimidine ring, keeping this proherbicide in its inactive form [6]. This latter reaction is analogous to the metabolic inactivation of chlorimuron ethyl by soybeans [87].

4 Resistance to Sulfonylurea Herbicides in Weeds

Resistance in weeds arising from treatment with sulfonylurea herbicides was first detected in 1987 [88, 89], five years after the first widespread commercial use of chlorsulfuron. Since that discovery, resistance to ALS inhibitors, including sulfonylureas and imidazolinones, has been documented in a total of 15 weed species (Table 5). These species include 11 dicotyledons, three monocotyledons and one sedge. The potential impact of resistance to ALS inhibitors on

Table 5. Resistance to ALS inhibitors

Weed Species	First Year Resistance Occurred (Non-Crop)	Selection Agent (a)	Location (State, Province, Country)	Total Sites to Date (Non-Crop)	Additional Reference(s)
Alopecurus myosuroides	1984	Chlorotoluron/ Isoproturon (b,e)	ENGLAND	1	
Amaranthus blitoides	(1991)	Sulfometuron methyl (c,e)	ISRAEL	(1)	
Amaranthus retroflexus	(1991)	Sulfometuron methyl (c)	ISRAEL	(1)	
Cyperus difformis	1992	Bensulfuron methyl	AUSTRALIA NSW UNITED STATES CA	9	DuPont Unpub.
Kochia scoparia	1987 (1987)	Chlorsulfuron (Sulfometuron methyl)	CANADA AB,MB,SK UNITED STATES CO,ID,KS,MT, NE,NM,ND,SD, OK,TX,WA	833 (62)	DuPont Unpub.
Lactuca serriola	1987 (1988)	Chlorsulfuron (Sulfometuron methyl)	UNITED STATES ID	10 (2)	
Lolium perenne	1991 (1989)	Chlorsulfuron (Sulfometuron methyl)	UNITED STATES CA,MS,TX	6 (2)	

Lolium rigidum	1989 1992	Chlorsulfuron Diclofop-methyl (b,e)	AUSTRALIA SPAIN	~100 (d) 2	 [120]
Sagittaria montevidensis calycina	1992	Bensulfuron methyl	UNITED STATES CA	35	DuPont Unpub.
Salsola iberica	1988 (1989)	Chlorsulfuron (Sulfometuron methyl)	CANADA SK UNITED STATES CA,WA,ND, MT,KS	52 (26)	DuPont Unpub.
Sinapis arvensis	1992	Chlorsulfuron	CANADA MB	1	DuPont Unpub.
Sisymbrium orientale	1991	Chlorsulfuron/ Triasulfuron (e)	AUSTRALIA SA	8	
Sonchus oleraceus	1991	Chlorsulfuron (e)	AUSTRALIA NSW	1	
Stellaria media	1988	Chlorsulfuron	CANADA AB DENMARK (one site)	29	DuPont Unpub.
Xanthium strumarium	1991	Imazaquin (f)	UNITED STATES MS,MO,TN	6	

Adapted from [90]. "Additional References" provided for updated information.

a. Unless otherwise stated, resistance mechanism to ALS-inhibiting herbicides is insensitive ALS.
b. Cross-resistant to chlorsulfuron.
c. Sulfometuron methyl and simazine were applied simultaneously but simazine probably dissipated before seed germination [90].
d. Estimates that up to 1000 sites may exist. Sites with non-ALS mechanism of resistance that are cross resistant to ALS inhibitors may increase this total.
e. Mechanism of resistance to ALS inhibitors is unknown.
f. Some sites are not cross-resistant to sulfonylureas.

agricultural production is substantial, and the subject has been reviewed periodically (see [90] for a detailed treatment) [91–93]. The following section summarizes the current agronomic status of resistance to ALS-inhibitor herbicides in weeds as well as recent advances in understanding fitness, gene flow, and mechanisms of resistance.

4.1 Occurrence of Resistance

Reliance on only ALS inhibitors for control of specific weeds in monoculture or non-crop areas (such as roadsides and railroad rights-of-way) has been the basis for most occurrences of resistance [88, 89, 94–99]. The use of ALS inhibitors either having long-residual weed control activity or being applied frequently (more than one treatment per season) also has contributed to the rapid onset of resistance [100]. These use patterns in conjunction with the very high efficacy of ALS inhibitors enhanced the selection intensity for resistant biotypes that already existed in wild populations at an estimated frequency of 1 in 10^6 [90, 100].

One of the first resistant *Kochia scoparia* sites, which was discovered in the western U.S. (Kansas) in 1987 [88], provides a typical example of this selection process. The field, in monoculture wheat production, had been treated with the longer-residual sulfonylurea chlorsulfuron five times during a five-year period [88, 101]. A resistant biotype(s) of the normally well-controlled *K. scoparia*

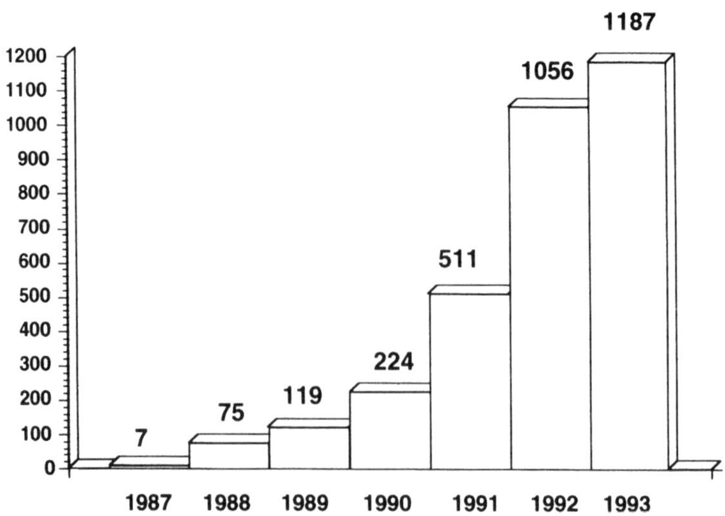

Fig. 6. Cumulative number of sites with documented resistance to ALS-inhibitor herbicides, as of October 1993

survived these treatments and proliferated until it dominated the weed population, resulting in weed control failure. This biotype exhibits a very high level of resistance. In greenhouse tests, the postemergent chlorsulfuron rate required for 50% growth inhibition was 30- to 431-fold higher for the resistant than a susceptible biotype, depending on the measurement used [88, 101]. Clearly this biotype could no longer be controlled in the field, and alternative control measures were required (see section 4.4 on resistance management strategies).

The number of sites with resistance to ALS inhibitors has increased steadily since the first site was discovered, and as of October 1993 there were nearly 1200 confirmed cases of resistance (Fig. 6). The majority of the sites have been documented in crop fields, while less than 10% have been reported in non-crop situations (Table 5). *K. scoparia*, which developed in non-crop areas as well as monoculture cereal fields in North America, accounts for approximately 75% of the resistant sites. However, just as the number of resistant sites has continued to increase since the first observations, so has the diversity of resistant species, the crops involved and the geographic regions affected (Table 5).

An important new development is the occurrence of ALS-inhibitor-resistant aquatic weed species biotypes in monoculture rice production in the U.S. (*Sagittaria montevidensis calycina* and *Cyperus difformis*) [94] and Australia (*C. difformis*) (Pappas-Fader unpublished data). The affected fields were monoculture rice fields that had received four annual applications of bensulfuron methyl. The U.S. resistant biotypes were not affected by typical application rates of bensulfuron methyl in the field, and the level of resistance to this herbicide ranged from 600- to 670-fold for *S. montevidensis* and 31- to 310-fold for *C. difformis* in greenhouse tests [94].

Also of concern is the relatively recent appearance of resistant *Xanthium strumarium* in monoculture soybeans in the southern U.S. Resistance in *X. strumarium* resulted from selection by repeated applications of imazaquin, a member of the imidazolinone class of ALS inhibitors [102, 103]. Normally *X. strumarium* is highly sensitive to imazaquin, and affected fields had received an annual preemergence imazaquin application that was sometimes followed by up to two postemergence applications. That resistance developed after only three to four years in a species whose other characteristics seem not to favor rapid development of resistance [100] is evidence for the overriding importance of selection pressure in promoting resistance. While some of the imidazolinone-resistant *X. strumarium* biotypes are not cross-resistant to other ALS inhibitors [104], at least one biotype does exhibit cross-resistance to sulfonylureas [103].

4.2 Mechanisms of Resistance

4.2.1 Target-Site Resistance

For a large majority of the weed biotypes with ALS-inhibitor resistance, the basis of resistance is reduced sensitivity of the target site acetolactate synthase (ALS) [90, 94, 95, 97, 98, 101]. Extensive information exists on the mutations of the ALS gene resulting in a herbicide resistant phenotype for a number of organisms including bacteria, yeast, and plants (see [90] for review). However, relatively little information is available on mutations in resistant weeds. Recently Guttieri et al. [105] demonstrated that a single point mutation in the codon for proline 173 in Domain A of ALS resulted in resistance in *Kochia scoparia* and *Lactuca serriola* biotypes. Domain A is a highly-conserved region of the protein in which a number of amino acid changes have been associated with resistance [106]. These researchers also reported that another resistant *K. scoparia* biotype and a resistant *Salsola iberica* biotype did not have mutations in Domain A, suggesting that other mutations in the gene for ALS may confer resistance to weeds selected in the field with sulfonylureas.

4.2.2 Cross-Resistance

Having target site insensitivity as the mechanism of resistance suggests that a resistant biotype selected by one ALS inhibitor may be cross-resistant to other ALS inhibitors, and in fact this has often been demonstrated [88–90, 94–98, 101, 103]. In general, resistant biotypes selected by a sulfonylurea also exhibit, to varying degrees, cross-resistance to other sulfonylureas as well as to triazolopyrimidines (another class of ALS inhibitors) and, frequently, imidazolinones. For example, resistant *K. scoparia* that was selected by several chlorsulfuron applications was 17-, 10-, 75-, 20-, and 9-fold less sensitive to chlorsulfuron, sulfometuron methyl, thifensulfuron methyl, triasulfuron, and imazapyr (an imidazolinone), respectively, than a susceptible biotype [88]. Similarly, resistant:susceptible ALS I_{50} ratios for the same biotypes also varied with the ALS inhibitor, ranging from 28 for sulfometuron methyl to 6 for imazapyr [101]. In contrast, as mentioned above, resistant *X. strumarium* biotypes selected by an imidazolinone and having ALS insensitive to this herbicide class so far have only occasionally been cross-resistant to other ALS inhibitors [103, 104].

Another consequence of target-site resistance is that the resistant biotypes that have been studied are controlled similarly to susceptible biotypes by herbicides with non-ALS-inhibitor modes of action [88, 89, 98, 103]. This finding has important implications for agronomic management of resistant biotypes (see below).

4.2.3 Non-Target-Site Resistance

A notable exception to the prevalent mechanism of resistance, insensitive ALS, is found in *Lolium rigidum* biotypes in Australia. At least one *L. rigidum* biotype has been selected with annual applications of chlorsulfuron and is resistant due to insensitive ALS [96]. However, several hundred *L. rigidum* biotypes in Australia that developed resistance to diclofop-methyl through frequent diclofop-methyl treatments are cross-resistant to chlorsulfuron, triasulfuron and other wheat-selective herbicides [91]. Two of these biotypes have been studied, and while ALS sensitivity to chlorsulfuron was equivalent to susceptible biotypes, metabolic inactivation of chlorsulfuron was two- to four-fold faster in the cross-resistant biotypes [96, 107]. Another example of non-target based resistance is an *Alopecurus myosuroides* biotype in England that developed metabolism-based resistance to chlorotoluron and isoproturon [108]. This biotype is cross-resistant to chlorsulfuron, but the mechanism of cross-resistance has not been established.

4.3 Fitness and Gene Flow

In contrast to the reduced fitness (survival and reproductive capability) observed in triazine-resistant plants [91], results of studies of fitness components in sulfonylurea-resistant and -susceptible biotypes have generally suggested that fitness is similar in resistant and susceptible biotypes (see [90] for a review). For example, relative competitiveness (the response of one biotype to competition from the other biotype) (Thompson and Thill unpublished results), growth [109, 110], and seed production [109] are similar in resistant and susceptible *K. scoparia* biotypes. Competitiveness [111], seed production [112] and seed longevity [112] are similar in resistant and susceptible *L. serriola*. Furthermore, catalytic competency of ALS is similar in four resistant weed species [95, 101]. Equivalent fitness between resistant and susceptible biotypes implies that a population that has become resistant will not readily revert to susceptibility when the selection pressure is removed [113].

Movement of resistance genes from the original site of resistance may occur through pollen or seed movement. Target-site resistance to ALS inhibitors in weeds is inherited as a single nuclear trait with dominance or incomplete dominance [114–116]. Thus pollen could play a role in the spread of target-site resistance. Movement of pollen substantial distances from a resistant plant source (up to 62 m) and transfer of resistance by pollen has been documented for *K. scoparia* in the field [117, 118], and it is likely for other outcrossing species as well. Seed movement by a variety of mechanisms can also be responsible for the spread of resistance (target-site or non-target-site) to previously uninfested areas. Wind (*L. serriola*), water flow (*S. montevidensis calycina*), animals (*X. strumarium*), tumbling plants (*K. scoparia, S. iberica*), and farm machinery are among the means by which seeds may easily move. For example, tumbling

S. iberica plants moved up to 5.5 km in 45 days, releasing seeds along the way [90, 119].

4.4 Management Strategies

Because of the apparent high fitness of resistant biotypes and the effective mechanisms of gene flow via pollen and seed movement, the potential for resistance to spread once it has become established is generally high. Therefore it is imperative that strategies be implemented to minimize the risk of developing resistance and to manage it if it does occur. Practices that reduce selection pressure include rotating crops in conjunction with rotating herbicide modes of action, using mixtures or sequential treatments with different modes of action, using short-residual herbicides, and limiting the number of treatments per crop with the same mode of action for control of the same weed. These practices should be designed into an integrated weed control program that incorporates cultural practices along with chemical methods. For example, timely use of tillage or cultivation, use of weed-free seeds and competitive crop stands, judicious use of non-selective herbicides, and preventing weed escapes from going to seed can each contribute to an effective resistance management program.

The importance of adopting effective resistance management strategies is evidenced by the proliferation of organizations of industry and/or academic scientists focused on this objective. One of the goals of international groups such as the ALS Inhibitor Resistance Working Group and the International Organization of Pest Resistance Management, as well as numerous regional and local organizations, is to support the development and use of such strategies through research and educational programs. In some cases, ALS-inhibitor herbicide labels now provide information on resistant weeds, including recommendations for resistance management.

Populations of several weed species have evolved the ability to survive sulfonylurea treatments through selection for resistance mechanisms. Resistance to sulfonylureas and other ALS-inihibitor herbicides is increasing in the number of sites, species, and geographies affected. Therefore, it is essential that management strategies continue to be developed and adopted to preserve the usefulness of these herbicides.

5 Concluding Remarks

Since their discovery in 1975, the development of the sulfonylurea herbicides by several companies has provided a wealth of new crop protection tools as well as new insights into herbicide biology, biochemistry and agrichemical behavior in the environment. To date, 17 distinct sulfonylurea herbicide active ingredients have been commercialized, and it is reasonable to expect that several others will appear during the 1990s. The properties of these herbicides and the history of their development offer an excellent example of what is possible in modern pesticide research. The sulfonylurea herbicides provide an immediate 10- to 100-fold reduction in application rates over existing standards, which translates to the dual positive consequences of far less compound applied to field soil and sharply reduced consumption of material and energy and production of chemical waste during associated manufacturing processes. Further, these modern herbicides have exceptionally positive toxicological and environmental profiles. Along with some other recently developed areas of herbicidal chemistry, the sulfonylureas provide a model for meeting the increasingly exacting needs of growers, society, and the environment.

Acknowledgments. We are pleased to acknowledge Elmo M. Beyer, Jon S. Claus and Kenneth A. Peeples for their valuable technical reviews and Sonia B. Combs for her skillful production of the tables, figures and final formatted manuscript.

6 References

1. Beyer EM, Duffy MJ, Hay JV, Schlueter DD (1987) In: Kearney PC, Kaufman DD (eds) Herbicides – chemistry, degradation and mode of action. Marcel Dekker, New York 3: 117
2. Blair AM, Martin TD (1988) Pestic Sci 22: 195
3. Beyer EM, Brown HM, Duffy MJ (1987) Proc Brit Crop Prot Conf Weeds 2: 531
4. Brown HM (1990) Pestic Sci 29: 263
5. Brown HM, Kearney PC (1991) In: Baker DR, Fenyes JG, Moberg WK (eds) Synthesis and chemistry of agrochemicals II. ACS symposium series 443, Washington, D.C., American Chemical Society p 32
6. Brown HM, Fuesler TP, Ray TB, Strachan SD (1991) In: Frehse H (ed) Pesticide Chemistry, XIV, Weinheim, Germany p 257
7. Levitt G (1991) In: Baker DR, Fenyes JG, Moberg WK (eds) Synthesis and chemistry of agrochemicals II. ACS symposium series 443, Washington, D.C., American Chemical Society p 16
8. Palm HL, Riggleman JD, Allison DA (1980) Proc Brit Crop Prot Conf – Weeds 1: 1

9. Upstone ME, Swann NB, Stichbury R (1982) Proc Brit Crop Prot Conf – Weeds 2: 629
10. Levitt G, Ploeg HL, Weigel RC Jr, Fitzgerald DJ (1981) J Agric Food Chem 29: 418
11. Doig RI, Carraro GA, McKinley ND (1983) Proc Brit Crop Prot Conf – Weeds 3: 1
12. E.I. du Pont de Nemours & Co., Bulletin No. AG-404, April 1984
13. Selley AP, Haney PJ, Austin JR (1985) Proc Brit Crop Prot Conf – Weeds 3: 931
14. Amrein J, Gerber HR (1985) Proc Brit Crop Prot Conf – Weeds 1: 55
15. Buchholz C, Davidson WE, Ahliker WL, LeRoy RL, Somody CN, Dorr JE (1986) Proc Western Soc Weed Sci 39: 139
16. Madsen K (1988) Weeds and weed control. 29th Swedish Weed Conf Uppsala p 90
17. Ferguson DT, Schehl SE, Hageman LH, Lepone GE (1985) Brit Crop Prot Conf – Weeds 1: 43
18. E.I. du Pont de Nemours & Co., Bulletin No. AG-1336 8036
19. Sionis SD, Drobny HG, Lefebvre P, Upstone ME (1985) Brit Crop Prot Conf – Weeds 1: 49
20. E.I. du Pont de Nemours & Co., Bulletin No E-62932, February 1985
21. Hoechst Aktiengesellschaft (1990) Technical Information HOE 075032 Experimental Herbicide
22. Hanna WS, Saari LL, E.I. du Pont de Nemours and Co. (personal communication)
23. Smith AE, Aubin AJ (1992) J Agric Food Chem 40: 2500
24. Nawamaki T, Suzuki K, Watanabe S, Yamamoto S, Kakuta K, Morimoto K, Oya E (1991) Proc Brit Crop Prot Conf – Weeds 1: 45
25. Takeda S, Yuyama T, Ackerson RC, Weigel RC, Sauers RF, Neal LW, Gibian DG, Tseng PK (1985) Weed Res (Japan) 30: 30
26. Yuyama T, Takeda S, Watanabe H, Asami T, Peudpaichit S, Malassa JL, Heiss P (1983) Proc 9th Asian-Pacific Weed Sci Soc Conf (suppl) p 554
27. Ackerson RC, Davis LA (1987) Proc 11th Asian-Pacific Weed Sci Soc p 137
28. Heiss PH (1987) Proc 11th Asian-Pacific Weed Sci Soc p 45
29. Takeda S, Erbes DL, Sweetser PB, Hay JV, Yuyama T (1986) Weed Res (Japan) 31: 157
30. Calderon JI, Hare CJ, Palis FV, Burhan H, Bhandhufalck A, Chong WC (1987) Proc 11th Asian-Pacific Weed Sci Soc p 73
31. Quadranti M, Rufener J, Zoschke A (1987) Proc 11th Asian-Pacific Weed Sci Soc p 117
32. Burhan H, Sozz D, Zoschke A (1989) Proc 12th Asian-Pacific Weed Sci Soc p 127
33. Zoschke A, Yun SK, Kiessling U (1989) Proc 12th Asian-Pacific Weed Sci Soc p 245
34. Suzuki K, Nawamake T, Watanabe S, Ikai T (1987) Proc 11th Asian-Pacific Weed Sci Soc p 461
35. Suzuki K, Watanabe S, Shirai Y, Endo T, Hirata H (1989) Proc 12th Asian-Pacific Weed Sci Soc p 141
36. Nissan Chemical Industries, Ltd. (1987) Experimental Herbicide NC-311
37. Suzuki K, Shirai Y, Hirata H (1990) In: Grayson BT, Green MB, Copping LG (eds) Pest management in rice. Elsevier, London, p 338
38. Asgrow Publications (1990) 104: 17
39. Ishida Y, Ohta K, Nakahama T, Yoshikawa H (1987) Eur Pat Appl EP 238070
40. Miki H, Ishida Y (1989) Eur Pat Appl EP 305939
41. Ishida Y, Ohta K, Nakahama T, Itoh S, Miki H, Yamada J, Kando Y, Masumoto K, Kamikado T, Yoshikawa H (1990) 7th Int Congr Pestic Chem (IUPAC) Hamburg. 01A-46
42. Ishida Y, Ohta K, Itoh S, Nakahama T, Miki H, Yoshikawa H (1993) J Pestic Sci 18: 175
43. Palm HL, Liang PH, Fuesler TP, Leek GL, Strachan SD, Wittenbach VA (1989) Proc Brit Crop Prot Conf – Weeds 1: 23

44. E.I. du Pont de Nemours & Co. (1989) DPX-E9636: Experimental herbicide for corn
45. Suzuki K, Nawamaki T, Watanabe S, Yamamoto S, Sato T, Morimoto K, Wells BH (1991) Proc Brit Crop Prot Conf – Weeds 1: 31
46. Slack CH, Witt WW (1992) Weed Sci Soc Amer Abstracts 32: 7
47. Ricotta JA, Bradshaw LD (1992) Proc N Central Weed Sci Soc 47: 46
48. Murphy GP, Witt WW (1992) Proc N Central Weed Sci Soc 47: 40
49. Bussler BH, White RH, Williams EL (1991) Proc Brit Crop Prot Conf – Weeds 1: 39
50. Flint JL, Miller KJ, Oyarzabal ES (1992) Proc N Central Weed Sci Soc 47: 47
51. Thompson PG, Flint JL, Sprankle PL (1992) Proc N Central Weed Sci Soc 47: 42
52. Rahn PR, Loussaert D (1992) Weed Sci Soc Amer Abstracts 32: 74
53. Ciba-Geigy Corp (1992) CGA-152005: Experimental herbicide for broadleaf weed control
54. Moseley CM, Abbott J, Cantwell J, Dill B, Johnson M, Kern C, Kidder D, Koenig J, Moses A, Pearson C, Pruss S, Schnappinger G, Stahlberg L, Stamm D, Threewitt T, Vitolo D, Zoschke A (1992) Proc N Central Weed Sci Soc 47: 38
55. Threewitt TB, Abbott J, Cantwell J, Dill B, Johnson M, Kern C, Kidder D, Koenig J, Moseley CM, Moses A, Pearson C, Pruss S, Schnappinger G, Stahlberg L, Stamm D, Vitolo D, Zoschke A (1992) Proc N Central Weed Sci Soc 47: 51
56. Hutchison JM, Peter CJ, Amuti KS, Hageman LH, Roy GA (1987) Proc Brit Crop Prot Conf – Weeds 1: 63
57. Parsons IM (1987) Proc Brit Crop Prot Conf – Weeds 3: 809
58. Thomas P (1984) Canola Growers Manual, Canola Council of Canada
59. Swanton CJ, Chandler K (1989) Can J Plant Sci 69: 889
60. Brown HM, Dietrich RF, Kenyon WH, Lichtner FT (1991) Proc Brit Crop Prot Conf – Weeds 2: 847
61. Lichtner FT, Dietrich RF, Brown HM (1994) manuscript in preparation
62. Hall JC, Swanton CJ, Devine MD (1992) Pest Biochem Physiol 42: 188
63. Buchanan FS, Swanton CJ, Gillespie TJ (1990) Can J Plant Sci 70: 873
64. Blackshaw RE, Derksen DA (1992) Can J Plant Sci 72: 203
65. Buchanan FS, Swanton CJ, Gillespie TJ (1990) Weed Sci 38: 389
66. Peeples KA, Moon MP, Lichtner FT, Wittenbach VA, Carski TH, Woodward MD (1991) Proc Brit Crop Prot Conf – Weeds 1: 25
67. Reese KD, Loux MM (1992) Proc N Central Weed Sci Soc 47: 124
68. Dexter AG, Luecke JL, Bredehoeft MW (1992) Proc N Central Weed Sci Soc 47: 125
69. Renner KA, Crook TM (1992) Proc N Central Weed Sci Soc 47: 126
70. Haga T, Tsujii Y, Hayashi K, Kimura F, Sakashita N, Fujikawa K (1991) In: Baker DR, Fenyes JG, Moberg WK (eds) Synthesis and chemistry of agrochemicals II. ACS symposium series 443, Washington, D.C., American Chemical Society, p 107
71. Ishihara Sangyo Kaisha Ltd. (1990) SL-160 (Flazasulfuron): A new selective turf herbicide
72. Reinke H, Rosenzweig A, Claus J, Kreidi M, Chisholm C, Jensen P (1991) Proc Brit Crop Prot Conf – Weeds 1: 445
73. E.I. du Pont de Nemours & Co., Bulletin No. H-36644, January 1992
74. Powell GE, Renner KA (1992) Proc N Central Weed Sci Soc 47: 114
75. E.I. du Pont de Nemours & Co. (1990) Bulletin No. H-01258
76. Obrigawitch TT, Kenyon WH, Kuratle H (1990) Weed Sci 38: 45
77. Kimura F, Haga T, Sakashita N, Murai S, Fujikawa K (1989) Proc Brit Crop Prot Conf – Weeds 1: 29
78. Maurer W, Gerber HR, Rufener J (1987) Proc Brit Crop Prot Conf – Weeds 1: 41
79. Ciba-Geigy Corp. Beacon herbicide technical bulletin, January 1988

80. Fonne-Pfister R, Gaudin J, Kreuz K, Ramsteiner K, Ebert E (1990) Pestic Biochem Physiol 37: 165
81. Neighbors S, Privalle LS (1990) Pestic Biochem Physiol 37: 145
82. Sweetser PB, Schow GS, Hutchison JM (1982) Pestic Biochem Physiol 17: 18
83. Anderson JA, Preister TM, Shalaby LM (1989) J Ag Food Chem 37: 1429
84. Cotterman JC, Saari LL (1989) Weed Sci Soc Amer Abstracts 29: 73
85. Meyer AM, Muller F (1989) Proc Brit Crop Prot Conf – Weeds 2: 441
86. Frear DS, Swanson HR, Thalacker FW (1991) Pest Biochem Physiol 41: 274
87. Brown HM, Neighbors SM (1987) Pestic Biochem Physiol 29: 112
88. Primiani MM, Cotterman JC, Saari LL (1990) Weed Technol 4: 169
89. Mallory-Smith CA, Thill DC, Dial MJ (1990) Weed Technol 4: 163
90. Saari LL, Cotterman JC, Thill DC (1994) In: Powles S, Holtum J (eds) Herbicide resistance in plants: Biology and biochemistry. Lewis (in press)
91. Holt JS, Powles SB, Holtum JAM (1993) Annual review of plant physiology and plant molecular biology 44: 203
92. Shaner DL (1991) In: Caseley JC, Cussans GW, Atkin RK (eds) Herbicide resistance in weeds and crops. Butterworth-Heinemann, Oxford, UK, p 187
93. Thill DC, Mallory-Smith CA, Saari LL, Cotterman JC, Primiani MM, Saladini JL (1991) In: Caseley JC, Cussans GW, Atkin RK (eds) Herbicide resistance in weeds and crops. Butterworth-Heinemann, Oxford, UK, p 115
94. Pappas-Fader T, Cook JF, Butler T, Lana PJ, Hare J (1993) Proc West Weed Sci Soc p 76
95. Saari LL, Cotterman JC, Smith WF, Primiani MM (1992) Pestic Biochem Physiol 42: 110
96. Christopher JT, Powles SB, Holtum JAM (1992) Plant Physiol 100: 1909
97. Hall LM, Devine MD (1990) Plant Physiol 93: 962
98. Kudsk P, Mathiassen SK, Cotterman JC (1994) Weed Research (in press)
99. Rubin B, Sibony M, Benyamini Y, Danino Y (1992) Weed Sci Soc Amer Abstracts 32: 66
100. Gressel J, Segel LA (1982) In: LeBaron HM and Gressel J (eds) Herbicide resistance in plants. John Wiley, New York, p 325
101. Saari LL, Cotterman JC, Primiani MM (1990) Plant Physiol 93: 55
102. Barrentine WL (1994) Proc South Weed Sci Soc Vol 47 (in review)
103. Kendig JA, DeFelice MS (1994) Weed Sci Soc Amer Abstracts 34: 12
104. Barrentine WL (1994) Proc South Soybean Conf, Memphis, TN (in review)
105. Guttieri MJ, Eberlein CV, Mallory-Smith CA, Thill DC, Hoffman DL (1992) Weed Sci 40: 670
106. Lee KY, Townsend J, Tepperman J, Black M, Chui CF, Mazur B, Dunsmuir P, Bedbrook J (1988) EMBO J 7(5): 1241
107. Cotterman JC, Saari LL (1992) Pestic Biochem Physiol 43: 182
108. Kemp MS, Moss SR, Thomas TH (1990) In: Green MB, LeBaron HM, Moberg WK (eds) Managing resistance to agrochemicals: from fundamental research to practical strategies (Washington, D.C., American Chemical Society) p 376
109. Thompson CR, Thill DC Weed Sci Soc Amer Abstracts 32: 44
110. Christoffoletti PJ, Westra P Weed Sci Soc Amer Abstracts 32: 17
111. Alcocer-Ruthling M, Thill DC, Shafii B (1992) Weed Technol 6: 303
112. Alcocer-Ruthling M, Thill DC, Shafii B (1992) Weed Technol 6: 858
113. Maxwell BD, Roush ML, Radosevich SR (1990) Weed Technol 4: 2
114. Mallory-Smith CA, Thill DC, Dial MJ, Zemetra RS (1990) Weed Technol 4: 787
115. Mulugeta D, Fay PK, Dyer WE, Talbert LE (1991) Proc West Weed Sci Soc p 81
116. Barrentine WL, Kilen TC (1994) Proc South Weed Sci Soc Vol 47 (in review)
117. Stallings GP, Thill DC, Mallory-Smith CA (1993) Weed Sci Soc Amer Abstracts 33: 60

118. Mulugeta D, Fay PK, Dyer WE (1992) Weed Sci Soc Amer Abstracts 32: 16
119. Mallory-Smith CA, Thill DC, Stallings GP (1993) Proc Brit Crop Prot Conf – Weeds 2: 555
120. De Prado R, Medina MA, De Prado JL, Jorrin J (1993) In: 8th EWRS Symposium "Quantitative approaches in weed and herbicide research and their practical application", Braunschweig p 301

Recent Studies of Imidazolinone Herbicides and Related Compounds

David W. Ladner

American Cyanamid Agricultural Research Center, P.O. Box 400, Princeton, New Jersey 08543-0400, USA

Imidazolinone herbicides have become established as an important class of herbicides. This report is a review of the recent work in this area. Chemical studies have provided novel derivatives and novel routes to new and existing analogs, and entry into the related series of compounds, the sulfonylcarboxamides. The newer structures have added to the understanding of structure–activity relationships, both in whole plants and at the enzyme level. Mode of action studies have helped clarify the cause of plant death, and studies on uptake and translocation have provided insight into key requirements for activity. More detail concerning the metabolism in plants of the compounds is now known and the effect of the safener, naphthalic anhydride, has been studied. Crop hybrids, resistant to imidazolinones have been developed, leading to expanded use of these compounds. Widespread use of imidazolinones and other AHAS-inhibiting compounds has elicited concern about herbicide resistance, and recommendations for dealing with this problem have been made. Future work with this major herbicide class will likely focus on new crop uses, controlling environmental behavior, and gaining a deeper understanding of the mode of action.

Chemistry of Plant Protection, Vol. 10

1 Imidazolinones – Background

Imidazolinone herbicides, discovered in the 1970s and developed in the 1980s, have been the subject of several reviews [1, 2]. Since those publications, work has continued in this important class of highly successful herbicides. The aim of this report is to review the more recent work and provide a perspective with respect to the previous knowledge.

This chapter is divided into three sections. The first provides a background summary, the second reviews the recent chemical discoveries including a related class of compounds, the sulfonylcarboxamides, and the third describes the recent biological studies.

For convenience throughout this document, the most common imidazolinone radical, chemically known as 2-[4,5-dihydro-4-methyl-4-(1-methylethyl)-5-oxo-1*H*-imidazol-2-yl]-, and shown in structures **4**–**7**, has been frequently abbreviated **Im**.

1.1 Historical Review of Imidazolinones

The imidazolinones were discovered through chemical modification of a random screening herbicide lead, the phthalimide **1**. Certain analogs of this lead were plant growth regulants [3] and further modifications led to the tricyclic structure **2**, a compound with improved herbicidal activity. Modification of **2** eventually led to the first imidazolinone structure, **3** [4].

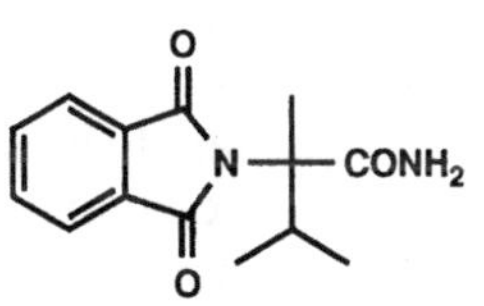

1

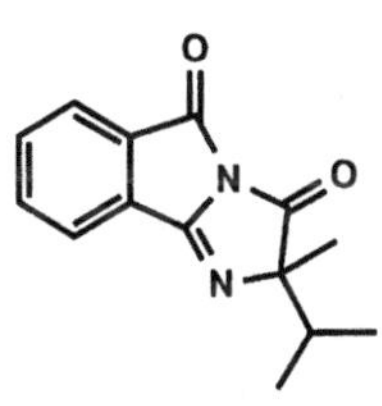

2

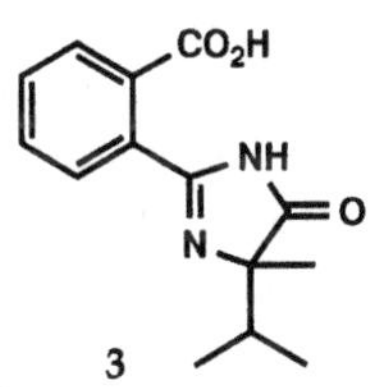

3

Analog synthesis of the benzene series produced a highly active pyridine compound imazapyr, 4, the 5-ethyl derivative imazethapyr, **5**, and the quinoline compound imazaquin, **6**. Together with imazamethabenz methyl, **7**, these four materials have been highly successful commercial products. Compounds **5** and **6** are used in soybeans, **7** is used in cereals, and **4** is a TVC agent, used in forestry, palm plantations and along rights of way [5].

The two soybean selective compounds, **5** and **6**, differ in weed spectrum; **5** is primarily a postemergence material in the central and northern US, and **6** is primarily preemergence in the southern United States and Latin America. Both are also sold as combination products with other herbicides. In addition, **6** is

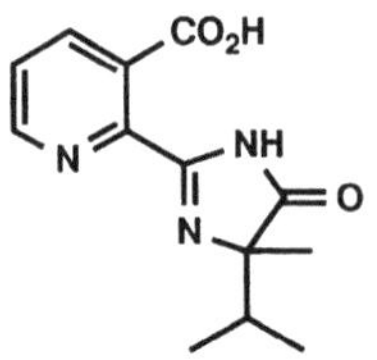

4
imazapyr

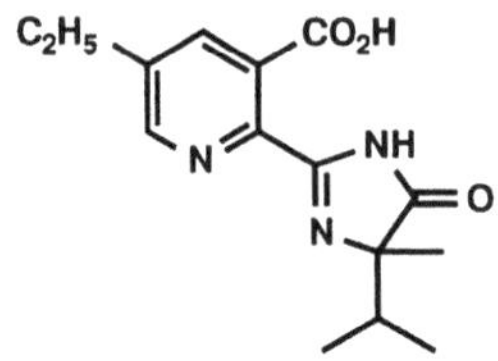

5
imazethapyr

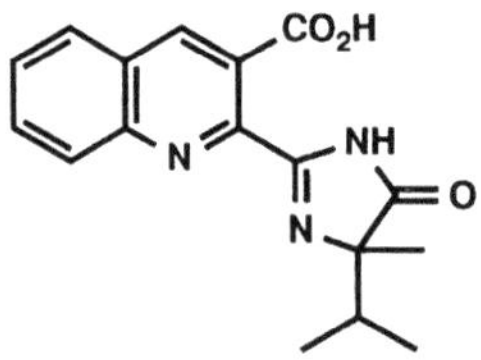

6
imazaquin

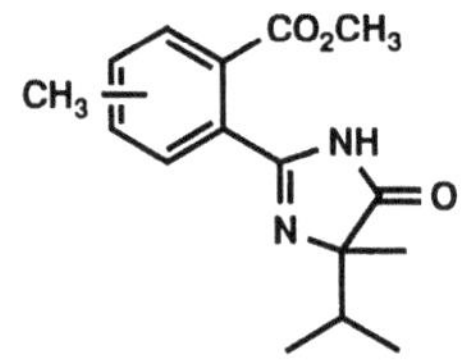

7
imazamethabenz methyl

used in southern United States turf for control of sedges (*Cyperus spp.*). Imazamethabenz methyl, **7**, is particularly effective in European wheat against mustard (*Brassicae spp.*) and wild oat (*Avena fatua*) [6].

The imidazolinones inhibit the production of the branched-chain amino acids and definitive experiments have pinpointed the mode of action as uncompetitive inhibition of acetohydroxyacid synthase (AHAS) [7]. Although there is structural similarity between the branched dialkylsubstituted imidazolinones and branched-chain amino acids, experiments have also shown that the herbicide binding site is not the amino acid-feedback regulatory site. Furthermore, kinetic studies and development of viable mutants resistant to imidazolinones have demonstrated that the compounds do not bind at the catalytic site. In this regard, the imidazolinones are similar to other AHAS inhibiting herbicides, although the exact binding location of each chemical class appears to be somewhat different [8] (see sect. 3.3).

Enzyme inhibition has been shown to be a requirement for activity, but not sufficient. The highly active herbicidal compounds translocate very efficiently to the growing meristematic tissue of susceptible species. This ability enhances their effectiveness in the field; despite lower enzyme-inhibiting activity than the sulfonylureas by a factor of 1000, field use rates of imidazolinones differ by much less, bringing some within a factor of 10 of the sulfonylureas [6].

Studies have unequivocally demonstrated that soybean and cereal selectivity is due to selective metabolic detoxification by the crop species. The metabolites depend not only on the compound but on the crop. Imazamethabenz methyl is selectively activated through de-esterification by susceptible weeds, but wheat preferentially detoxifies the compound by hydroxylating the methyl side chain. Earlier results also showed that imazethapyr is metabolized by soybean at the ethyl group [9]. Imazaquin, however, undergoes metabolism at the imidazolinone ring. Recent studies, reported below, have identified the nature of the metabolites.

A fundamental property of imidazolinones is low mammalian toxicity [10], an outcome of their plant-specific site of action. This target specificity has prompted general interest in amino acid biosynthesis pathways as herbicide sites of action [11].

2 Recent Areas of Imidazolinone Research

This section is devoted to the recent chemical studies in imidazolinone chemistry and a related series, the sulfonylcarboxamides. Although imidazolinones appear to have fairly strict structural requirements for herbicidal activity, recent patents and applications have revealed further variations of bioactive structures. Equally important has been the discovery of several new synthetic routes to commercial and novel compounds. The sulfonylcarboxamides represent a generically new class of compounds, but are included here because they were direct chemical descendants of imidazolinones and share some common structural features.

2.1 Recent Novel Chemistry of the Imidazolinones

The structural types of the imidazolinones which have appeared in recent disclosures or not previously discussed can be divided into five categories. These are as follows:

1) derivatives of the parent products as pro-herbicides, usually as a masked acid;

2) substituted imidazolinone ring analogs in which the isopropyl and methyl groups have been replaced;

3) N-substituted imidazolinones;

4) modified or replaced aromatic backbones; and

5) novel imidazolinyl 5- or 6-substituted pyridine carboxylic acids.

Each of these will be discussed in turn.

2.1.1 Acid Equivalents

A number of patents have appeared which have claimed derivatives of the parent carboxylic acid. These have included addition compounds of the known tricyclic analogs, **8** [12], and **9** [13], heterocycle-containing esters **10** and **11** [14, 15], oxime and aminoethyl esters **12** and **13** [16–18], six-membered tricyclic structures, **14** and **15** [19, 20], and condensation products, **16** and **17** [21, 22].

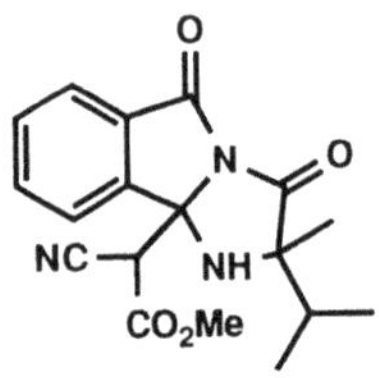

8

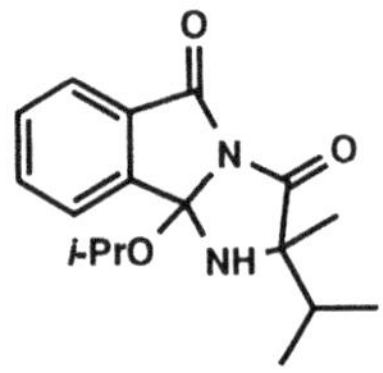

9

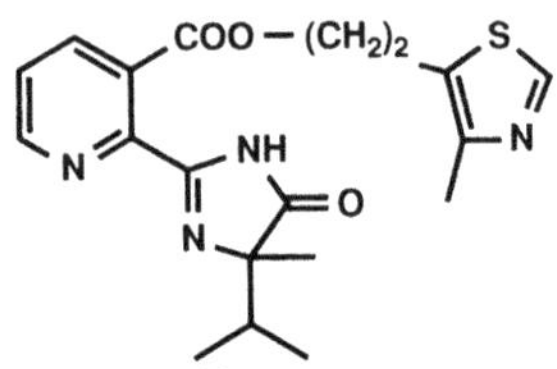

10

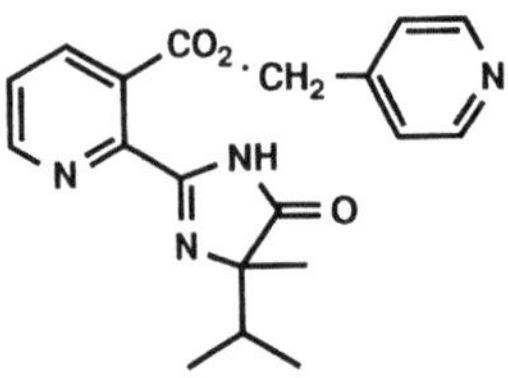

11

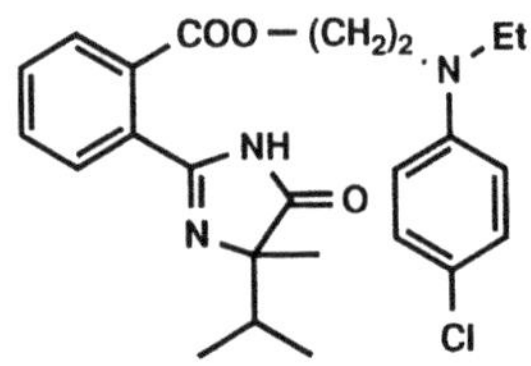

12

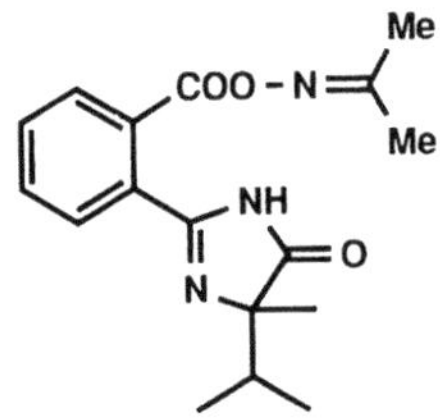

13

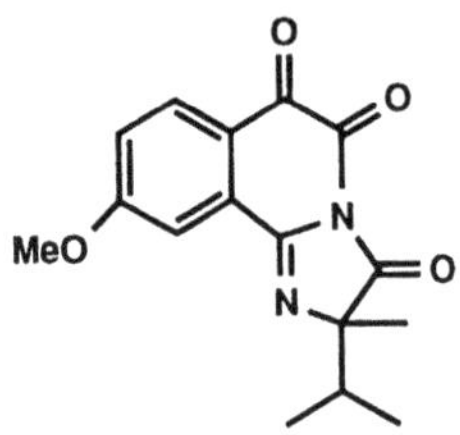

14

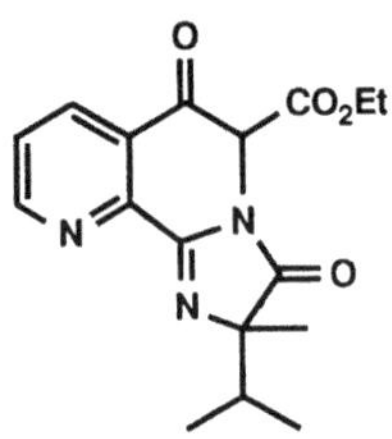

15

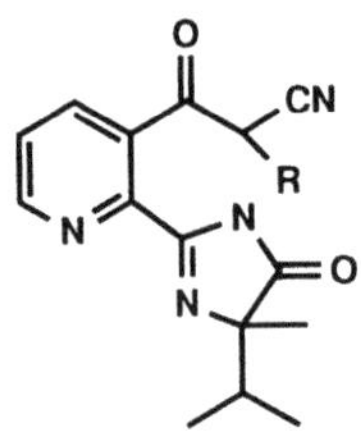

16, R= CO_2Me
17, R= heteroaryl

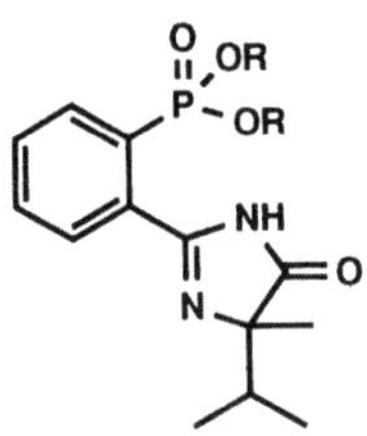

18

Presumably, these compounds are "pro-herbicides" and have the potential to metabolize differently from the parent acids. Since none of these materials have been commercially developed, however, little information on their field behavior is known. Finally, synthesis of a 2-phosphonic acid analog, **18**, has been described [23].

2.1.2 Imidazolinone Ring Substituents

Other recently disclosed analogs contain novel imidazolinone ring substituents in place of the familiar isopropyl, methyl groups. Examples of these are **19**, where

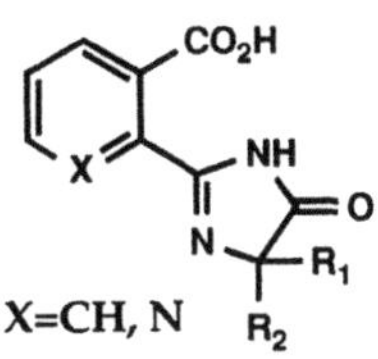

X=CH, N

19

R1	R2
$-CH_2CH_2SCH(CH_3)-$	
$-CH_2SCH_2CH(CH_3)-$	
CH_3	cyclohexen-4-yl
CH_3	cyclopenten-3-yl
CH_3	$-CH(CH_3)CH_2SCH_3$

R_1 and R_2 are cycloalkenyl, alkylthioalkyl, thiaspirocyclo [24], and fluoroalkyl groups [25].

2.1.3 *N*-Substituted Derivatives

Unlike *N*-alkyl and *N*-acyl compounds reported earlier, recently described *N*-substituted imidazolinones possess strongly electron-withdrawing substituents. These have the potential to undergo ring opening and deactivation to **21** in tolerant plants or soil:

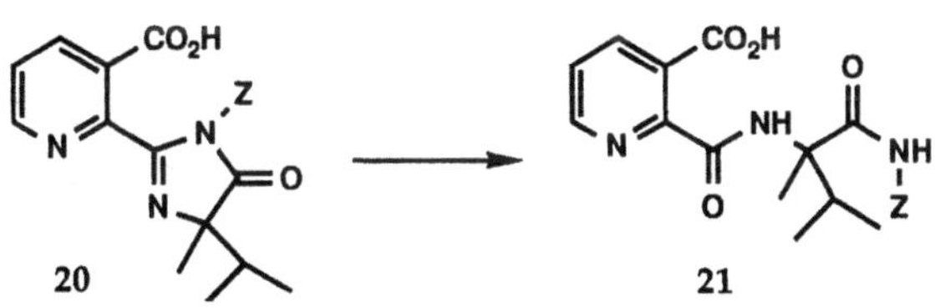

20 21

Z = electron-withdrawing group

Examples of these appear below. A more detailed account of this series has been published [26].

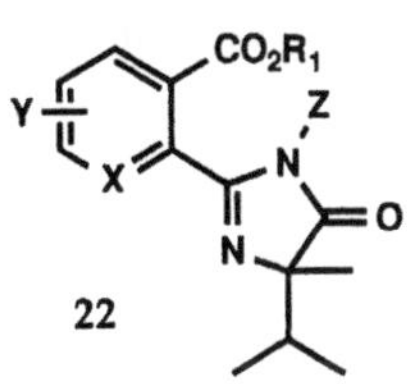

22

R_1= H, CH_3, furfuryl, p-MeO-benzyl, allyl, N^+Bu_4

X=CH, N

Z=Cl, OH, CN, SO_2Ar

Y=H, Me, Et

Some general routes to these compounds are shown in Scheme 1.

The *N*-tosyl compounds were a special case of this compound class. The hydrolysis of **29** (Ar = *p*-tosyl) to the ring-opened compound **30** occurred quite easily. Contrary to the expected loss of activity, **30** was herbicidal in its own right, and this led to the investigation of a new class of AHAS inhibiting herbicides, the

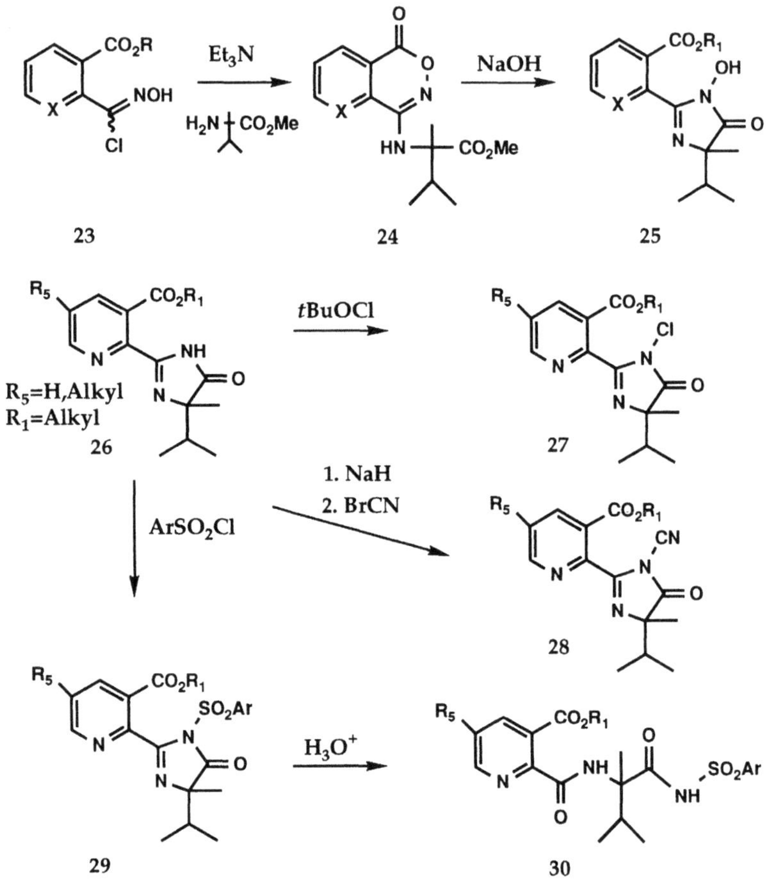

Scheme 1. Routes to N-substituted imidazolinones

sulfonylcarboxamides. The chemistry of these will be discussed in more detail in sect. 2.2.

2.1.4 Modifications of the Aromatic Ring Backbone

Variations on the aromatic ring have also been the subject of some continued interest. In addition to the disclosures on heteropyridines reported earlier [27, 28], a series of highly active heterobenzenes has emerged [29–31]. The most important of these are structures **31**–**34**.

These compounds were prepared from benzene diesters, diacids or cyclic anhydrides by conventional means [1, 32], although a mixture of both angularly fused regioisomers were usually formed. Separation was required since the

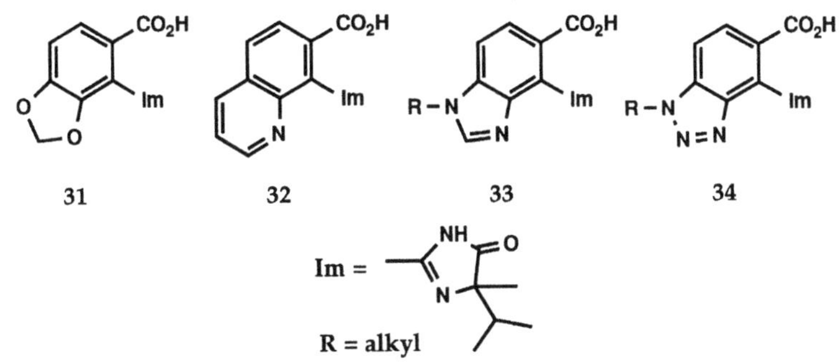

isomers with the heterocyclic ring fused next to the carboxylic acid were much less active. The linearly fused compounds were likewise less interesting. The diester intermediates were often the most difficult synthetic problem. Some synthetic routes to these compounds are shown in Scheme 2. The major obstacle was lack of regiospecific control at a number of key steps, which often resulted in isomeric mixtures and tedious separations.

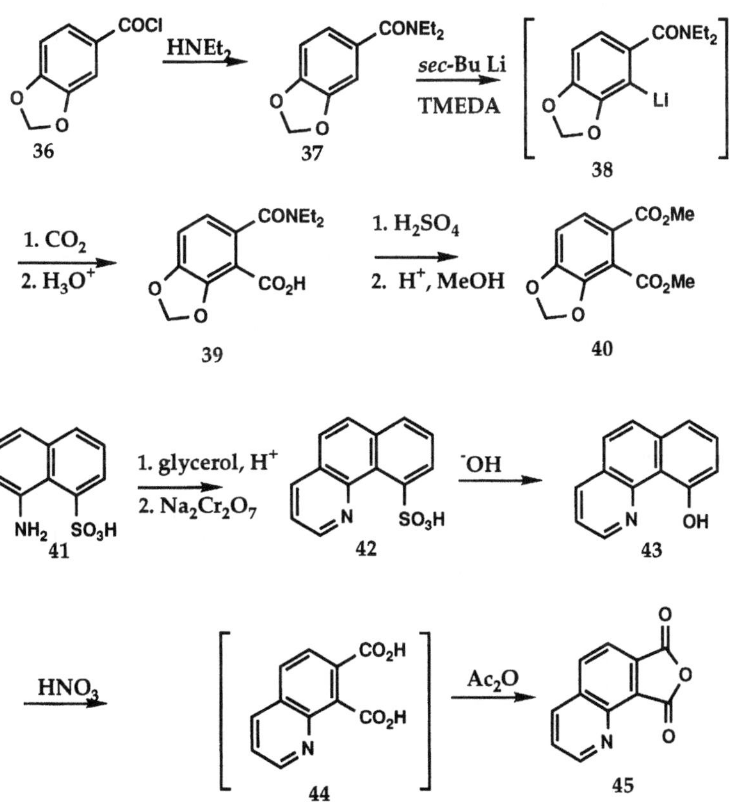

Scheme 2. Synthetic routes to heterobenzene diesters and anhydrides

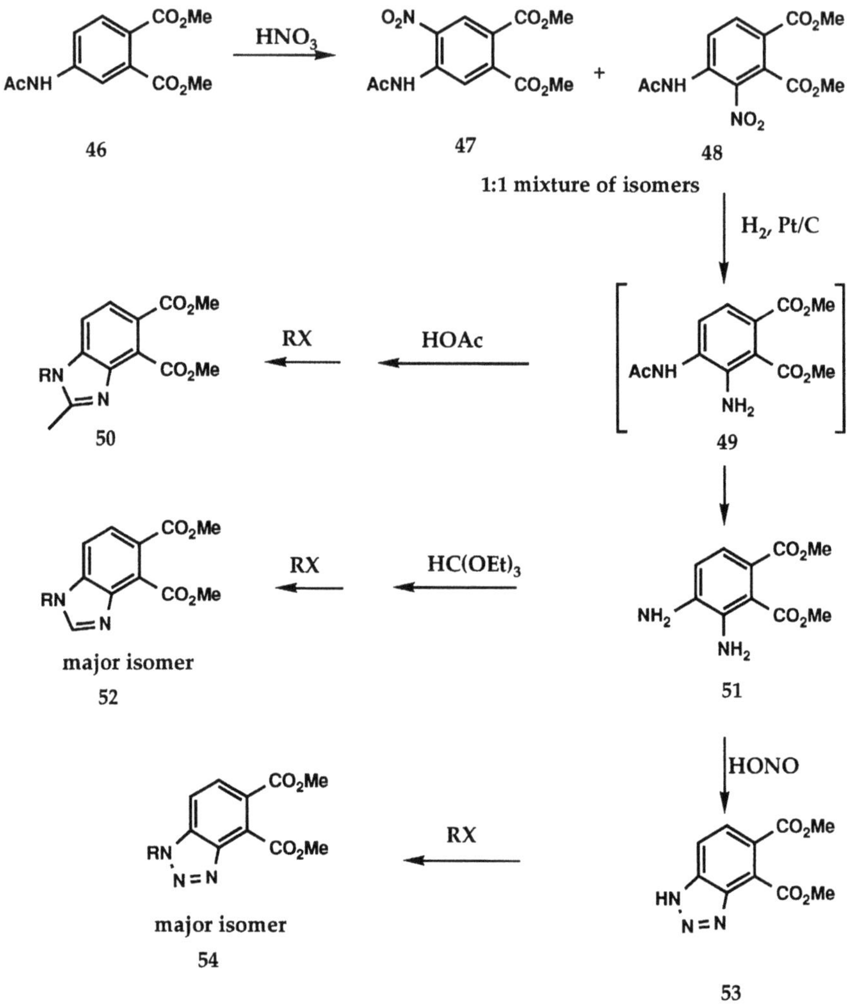

Scheme 2, continued

Nissan as well as Ciba-Geigy have also disclosed di- and tetrahydro quinolines [33, 34], compounds **55** and **56**, respectively.

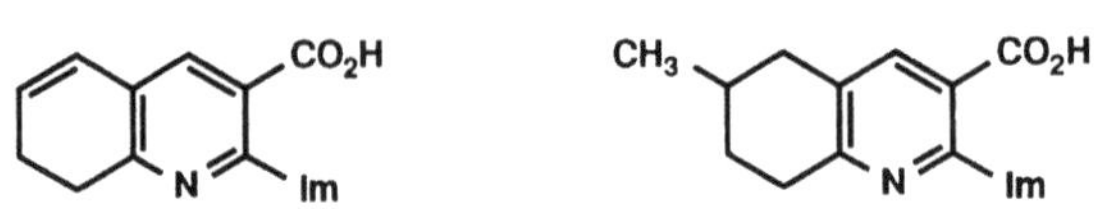

55 56

There have been a number of unusual backbones claimed in the patent literature which depart more significantly from the usual highly active structures.

Mitsui Toatsu [35] has published a thiazole backzone **57** and Bayer [36] has recently claimed the pyrimidines **58**, although both of these backbones had been disclosed earlier [2].

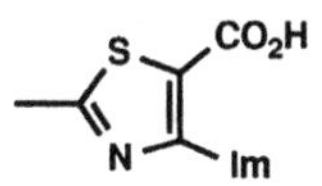

57

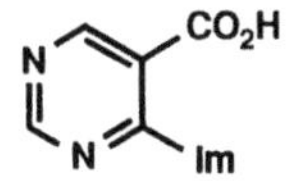

58

Patents continue to appear in the area of non-aromatic backbones. In addition to the original 2,3-dimethyl maleic acid-based structures published by three companies [37], two somewhat different series have appeared, **59** and **60** [38, 39].

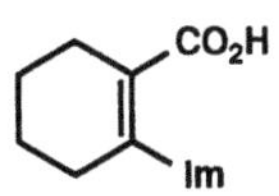

59

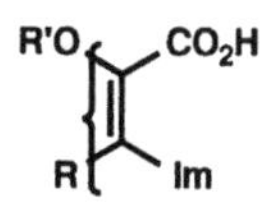

60

R, R' = C_1-C_6 alkyl

A curious variation, **61**, was disclosed some time ago [40] and is included here for completeness. The structure is technically an imidazolinone, but is probably not an AHAS inhibitor. The imidazolinone moiety in this case is probably acting as a masked carboxylic acid in a phenoxypropionic acid-type herbicide.

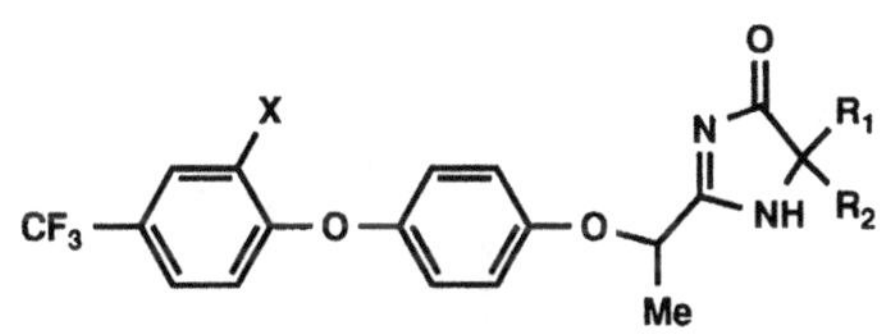

61

2.1.5 Recent 5- and 6-Substituted Imidazolinylpyridinecarboxylic Acids

A substantial amount of imidazolinone chemistry has centered around the highly active pyridine nucleus, i.e., analogs of imazapyr, **4**, and imazethapyr, **5** [41]. Previous investigations have shown that the 5 and 6 positions of the pyridine ring can tolerate a fairly wide range of substituents, although highest activity falls within a narrower range. In the search for novel compounds attention was given to those with improved behavior, such as selectivity in crops other than soybeans and lower soil activity. As in other series, the availability of the diesters was the key to the preparation of imidazolinone final products. A representative list of these compounds appears in Table 1.

Table 1. 5-and 6-substituted imidazolinylpyridinecarboxylic acids

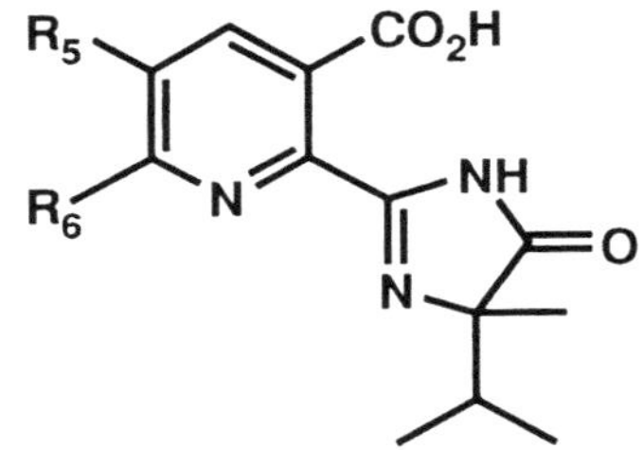

Compound	R_5 or R_6 Group[a]	Synthetic Route	Compound	R_5 or R_6 Group[a]	Synthetic Route
62	$(CH_3O)_2CH$-	A	**71**	formyl	A
63	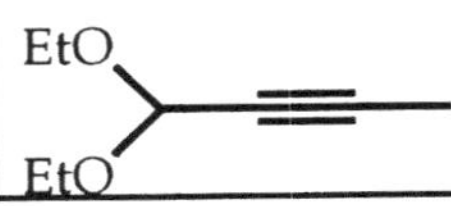	B	**72**	MeO_2C-C≡C-	B
64	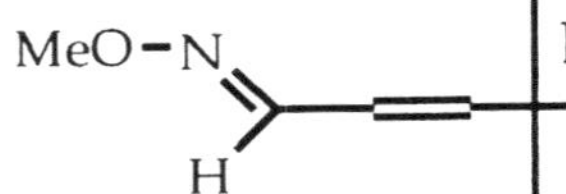	B	**73**	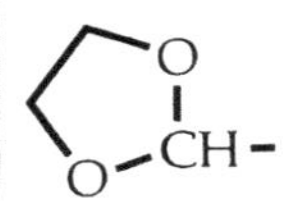	A

65	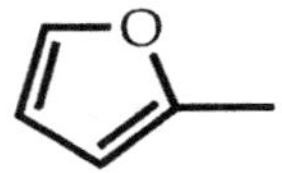	B	**74**	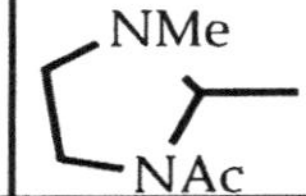	A
66	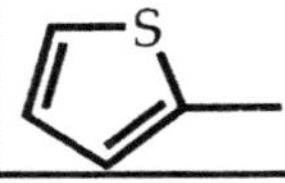	B	**75**	azido	D
67	$EtO_2CCH{=}CH$-	B	**76**	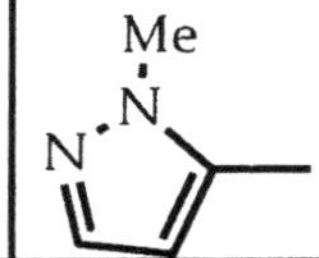	B
68	carbomethoxy	A	**77**	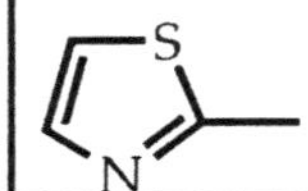	B
69	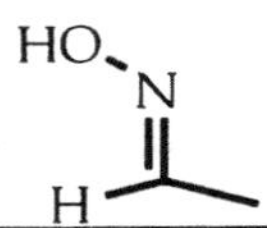	C	**78**	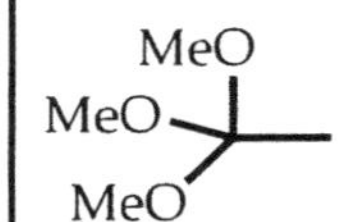	A
70	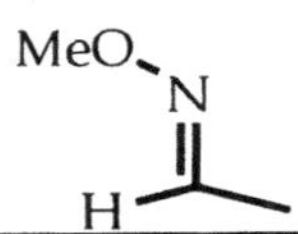	C	**79**	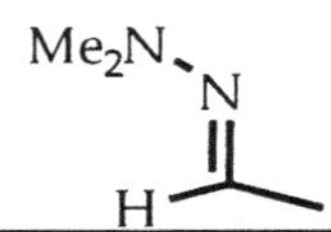	C

[a] 5- substituent unless otherwise designated

Table 1. Continued

Compound	R5 or R6 Group[a]	Synthetic Route	Compound	R5 or R6 Group[a]	Synthetic Route
80	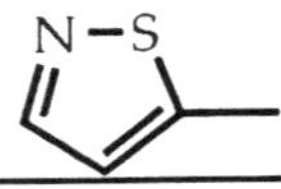	B	**87**	6-$PhCH_2S$-[42]	F
81	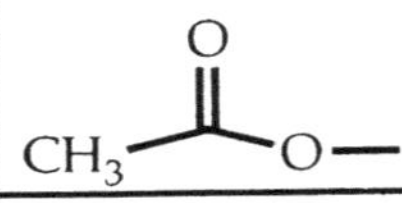	E	**88**	5-Et, 6-$MeOCH_2$-[43]	G
82	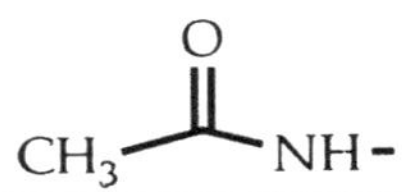	E	**89**	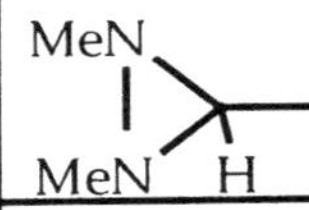	A
83	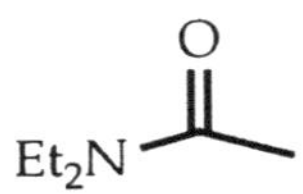	A	**90**	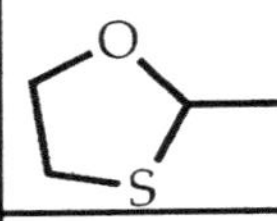	A
84	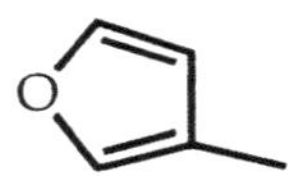	B	**91**	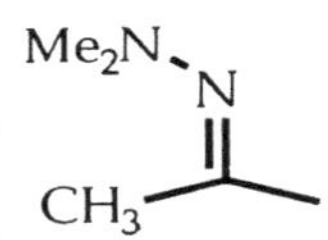	B[44]

85	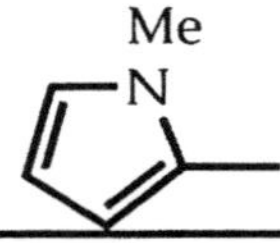	B	**92**	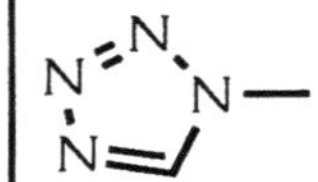	H
86	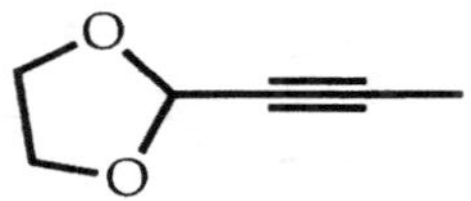	B	**93**	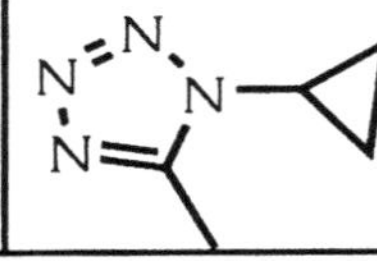	H

a 5- substituent unless otherwise designated

The key methods are as follows.

A. Displacement reactions of 5-mono-, di- or trihalomethylpyridine diesters with the appropriate nucleophile. The formed acetals, ketals or ortho esters could then be hydrolyzed to the carbonyl compound, in situ or in a separate step. In the conversion of triester **99** to the imidazolinone **101**, the one-step procedure [32] was not only regiospecific, but the 5-carbomethoxy group remained intact while the 3-carbomethoxy group was converted to the acid.

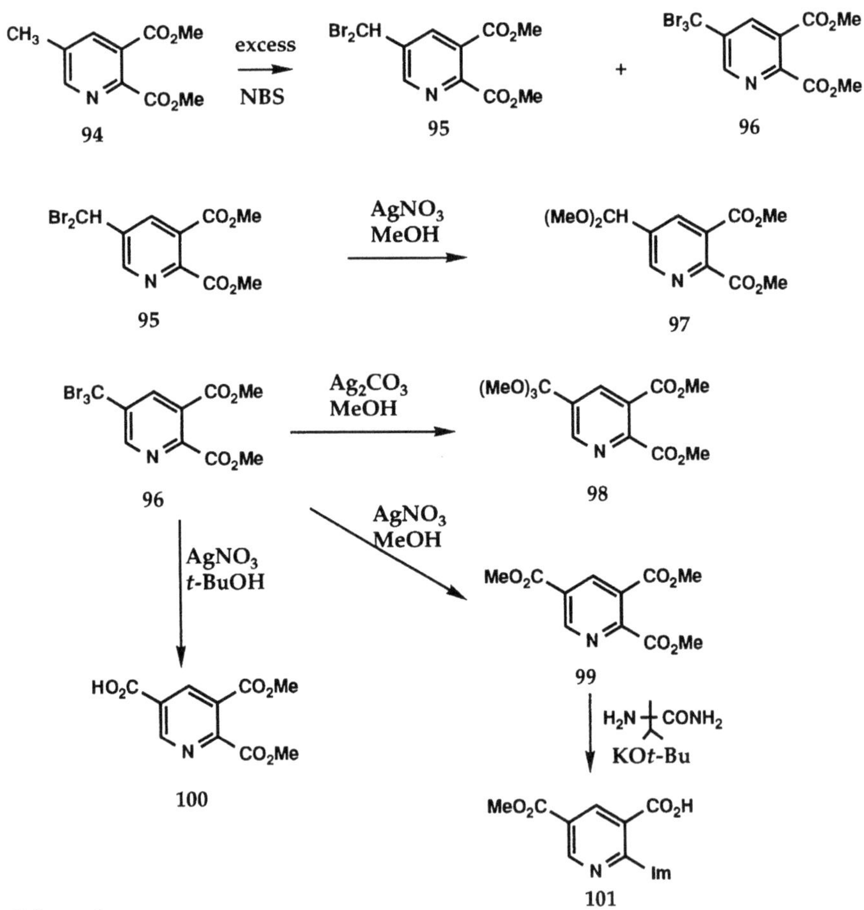

Scheme 3

B. Palladium-catalyzed coupling reactions with the 5-bromopyridine diesters, **102**.

This scheme [45] and modifications of it have provided a variety of novel substituents, particularly heterocyclic, alkenyl and alkynyl derivatives. The

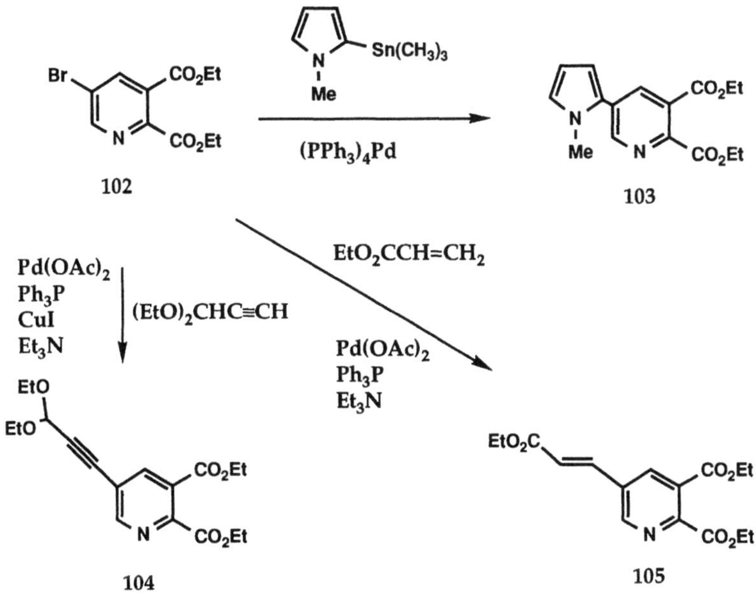

Scheme 4

remaining methods are straightforward manipulations of compounds previously described.

C. Condensation of the appropriate amino compound with the 5-acetyl or 5-formyl compound; the amino compound was obtained from either nitro reduction or Beckman rearrangement of the 5-acetyl derivatives.

D. Diazotization of the 5-amino compound and subsequent displacement with azide.

E. Acetylation of the 5-amino or 5-hydroxy compounds.

F. Displacement of the 6-chloro compound with an alkylthio anion.

G. A Minisci reaction of a 5-substituted pyridine diester with a carboxylic acid in the presence of Ag(II) and peroxysulfate giving a 5,6-disubstituted compound [44].

H. Reaction of the appropriate nucleophile (RNH_2 or N_3^-) with the corresponding imidoyl chloride. The imidoyl chloride was prepared by $SOCl_2$ treatment of the *N*-alkylcarbamoyl derivative.

2.1.6 Novel Synthetic Routes to Pyridine Diesters

Widespread acceptance of imazethapyr and the development of AC 263,222, has generated an interest in the diesters **94** and **106**. Several chemical firms have patented routes to these key intermediates. Four major synthetic routes which have been reviewed previously [46] are (1) the reaction of acroleins with chlorodiethyloxalacetate, (2) the oxidation of substituted quinolines, (3) condensations of enamines or active methylene compounds with ethoxymethylene diethyloxalacetate, and (4) the condensation of 2-aminomaleates with either propargyl aldehydes or 2-alkylpropenals followed by oxidation. Below are illustrated some recent innovations. The first are various cycloaddition approaches patented by Hoechst [47], **A**; ICI [48] (Zeneca), **B**; Shell [49], **C**; and Ciba-Geigy [50], **D**:

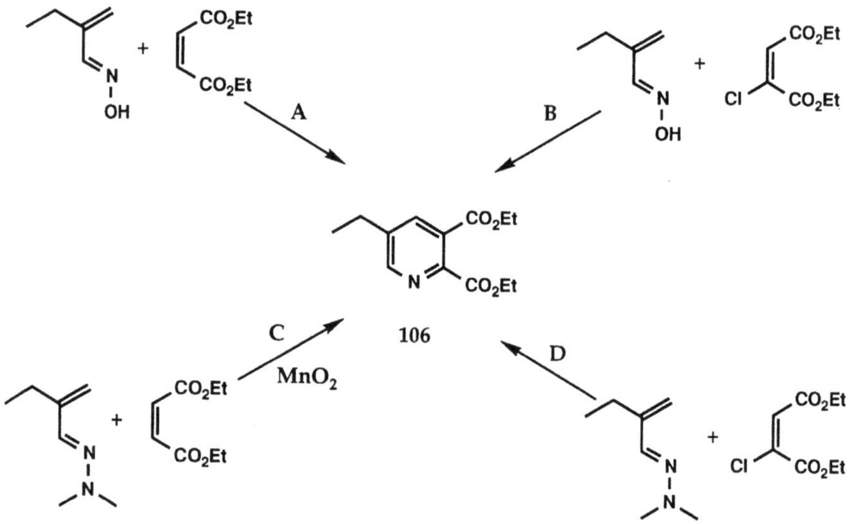

Others are the use of chloroaminomaleate, **107**, by Wacker [51], the hydroxylaminosuccinate, **108**, by Hoechst [52], and more recently, the hydroxydihydropyran, **109**, by Kuraray [53]:

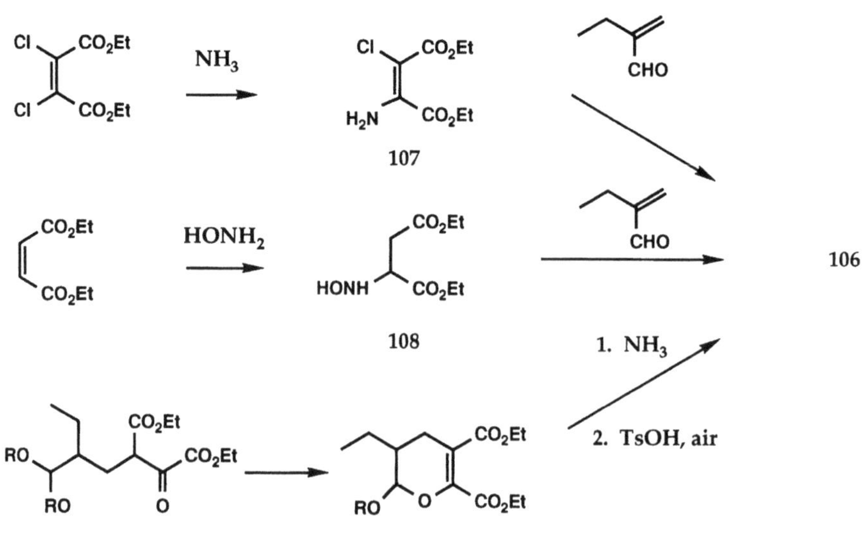

In the heterocyclic[*b*]pyridine series, two routes to diesters deserve special note. The first is an approach based on reactions reported in 1977 [54] and 1983 [55], then further described by Seitz [56] and Taylor [57] in which ω-alkynylthio- or oxo-1,2,4-triazenes undergo Diels-Alder/Retro Diels-Alder cyclo addition to provide nitrogen, oxygen or sulfur-containing heterocyclic fused pyridines. Synthesis of the dihydrofuro[2,3-*b*]pyridine diester, **112**, and the thiopyrano[3,4-*b*]pyridine diester, **113**, illustrate the utility of this approach [27].

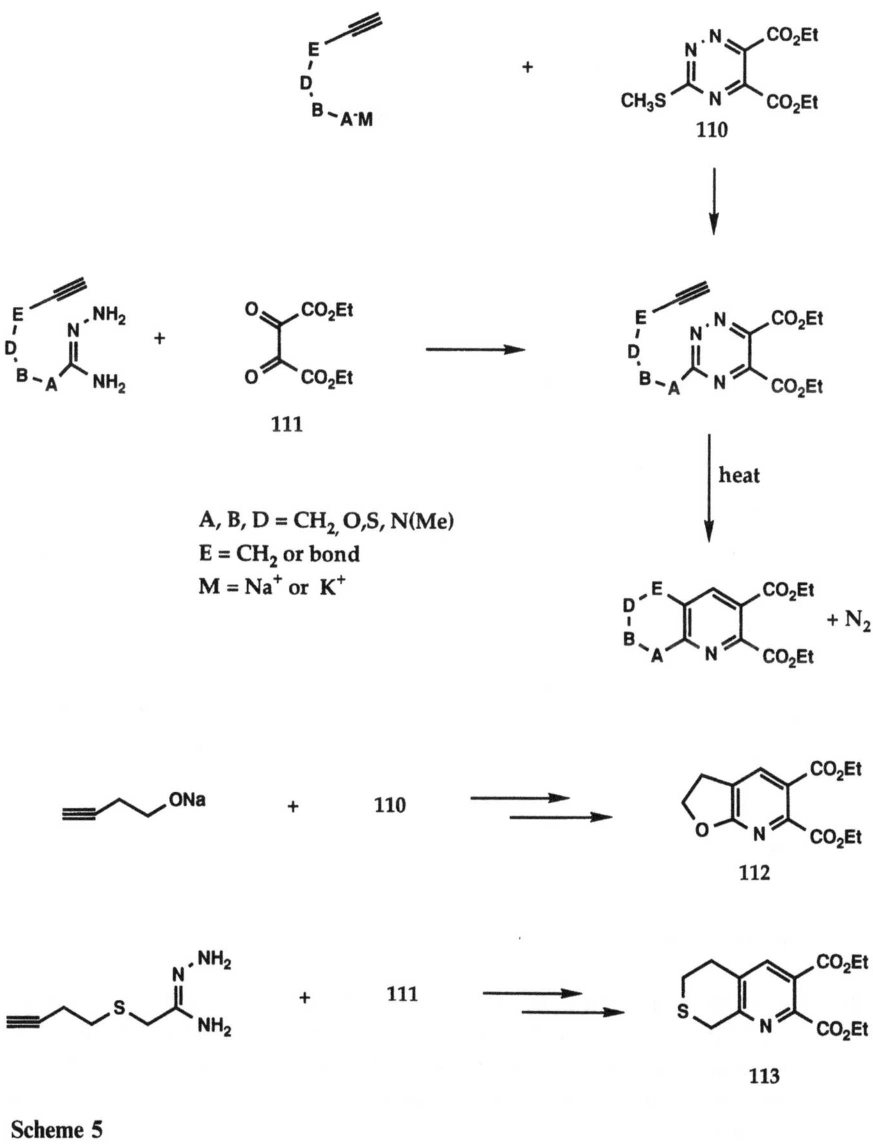

Scheme 5

Another useful approach to heteropyridines was the exploitation of the [2 + 2 + 2] cobalt- or rhodium-catalyzed cycloaddition reactions suggested by the work of Vollhardt [58] and Grigg [59]. This allowed access to a variety of oxygen, nitrogen or sulfur heterocyclic fused pyridines; some examples are shown below [58]:

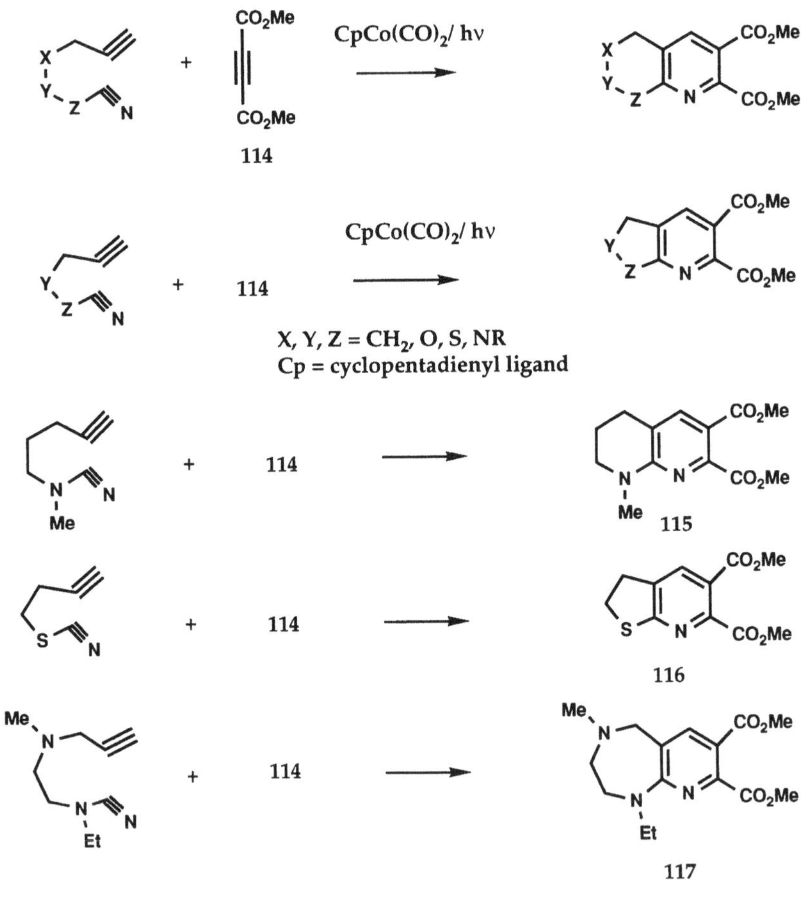

Scheme 6

2.2 Sulfonylcarboxamide Synthesis

In the synthesis program to examine *N*-substituted analogs, the *N*-tosyl compound **118** was prepared. While attempting to hydrolyze this material to the corresponding acid, the sulfonylcarboxamide, **119**, was isolated instead. This compound had unexpected herbicidal activity and was the basis for a new synthesis program.

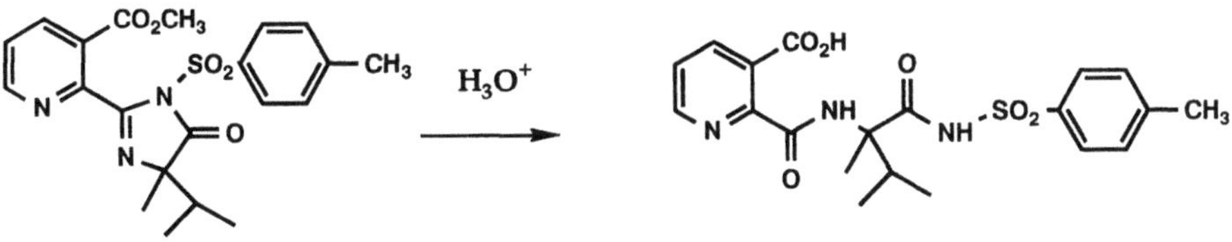

118 119

Several hundred analogs which incorporated the structural features of both imidazolinones and sulfonylureas were prepared. The most active were those represented by the generic structure **120**:

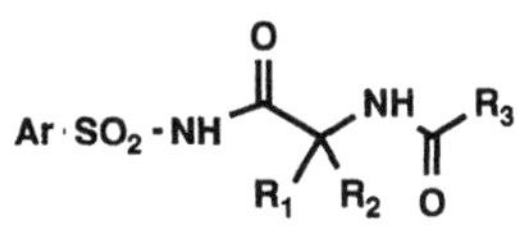

120

R_1, R_2, R_3 = alkyl, Ar = substituted phenyl

A detailed report on sulfonylcarboxamides has recently been published [60]. The chief synthetic routes to **120** were via ring opening of an imidazolinone, **121**, or an azlactone, **125**.

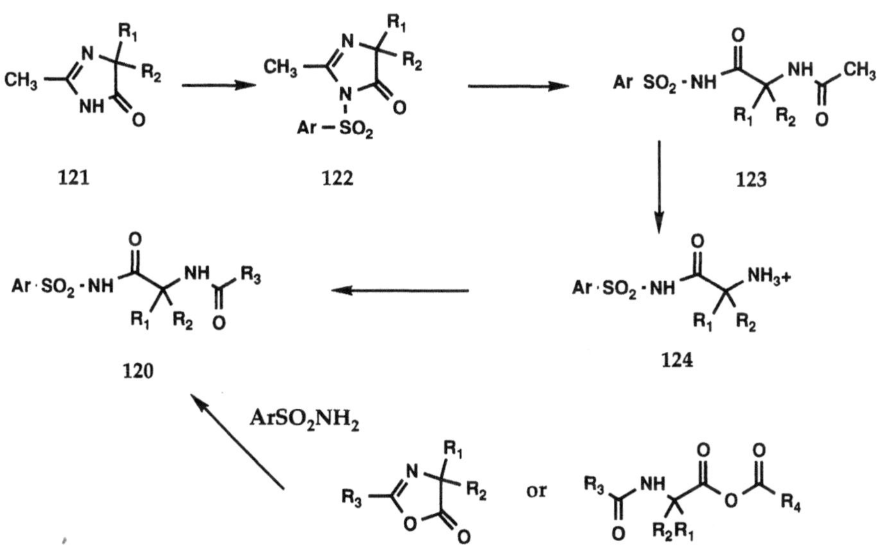

121 122 123 124 120 125 126

A number of related patents and publications have appeared containing these types of compounds [61–65]. The above generic for **120** was expanded to encompass compounds where Ar is phenyl, heteroaryl and alkyl. Also active were carbamates and ureas, i.e., compounds where R_3 is alkoxy or alkylamino. Like the imidazolinones, the compounds where R_1 and R_2 are isopropyl and methyl could be prepared as optical isomers, starting from the optically active amino amide, resolved in the previously described manner [1a].

3 Recent Biological Studies of Imidazolinone and Sulfonylcarboxamides

Work has continued to gain greater understanding of herbicidal structure–activity relationships, the mechanism of AHAS inhibition, uptake and translocation, and the nature of metabolism and selectivity. In addition, an issue of increasing importance for all AHAS-inhibiting compounds has been the study of resistance in crops and weeds.

3.1 Structure–Activity Relationships of Imidazolinones

The conclusions in previous reports regarding the effect of each structural domain on both herbicidal activity and AHAS activity generally remain intact. For discussion purposes, the structural domains of the imidazolinones are designated as the carboxylic acid, the imidazolinone ring, the aromatic backbone and the backbone substituents.

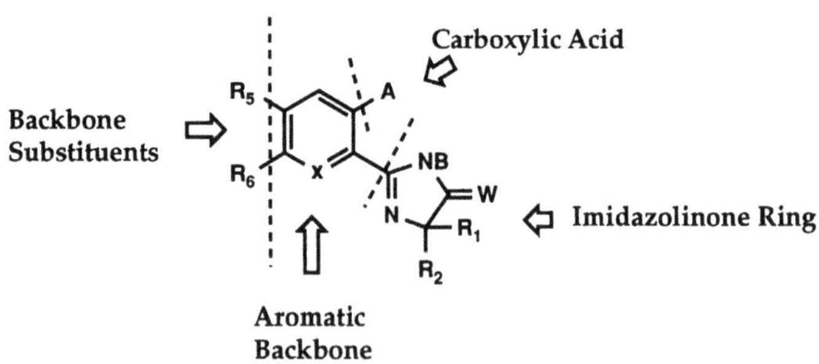

127

Replacement of the aromatic backbone by a 2-butenyl backbone has given active compounds [37–39], but the inherent lability of these types of compounds may hinder their utility as widely useful herbicides. Lability could be an asset to overcoming any soil persistence however, and different ways to introduce this property have also emerged. In particular, the *N*-substituted compounds and certain novel 5-substituted imidazolinylpyridines have shown this property. The major concern has been the search to find new highly potent compounds as well as those which are selective in major crops.

3.1.1 The Carboxylic Acid and Equivalents

None of the compounds mentioned in sect. 2.1.1 have shown superior activity to the parent acid, although labile esters approach the acid in potency. Useful selectivity has not been evident and there are no known developmental candidates in this area.

3.1.2 N-Substitution and R_1, R_2 Replacements on the Imidazolinone Ring

No improvement has been found for [methyl, isopropyl] as the substituents on the ring. The compounds, **19**, discussed in sect. 2.1.2, were substantially less active.

Quite good activity was found for some of the *N*-substituted compounds. Certain of these had selectivity in sunflower (*Helianthus annuus*): **25**, X = N, R_1 = H; soybeans (*Glycine max*): **27**, R_1, $R_5 = CH_3$ and **28**, $R_5 = C_2H_5$, R_1 = 2-furfuryl, 4-methoxybenzyl or Bu_4N^+; wheat (*Triticum aestivum*): **28**, $R_5 = C_2H_5$, R_1 = 2-furfuryl, 4-methoxybenzyl; corn (*Zea mays*) **28**, $R_5 = CH_3$, R_1 = allyl. There is evidence that the N-cyano compounds, **28**, break down rapidly in the soil. Compounds in this series have been discussed at length [27].

3.1.3 The Aromatic Backbone: Structure–Activity Profile of the Heterobenzenes

Probably the most striking result, at least based on previous experience, was the level of herbicidal activity obtained with certain heterobenzene compounds. The compounds **31**, **33** (R = Me) and **34** (R = Me) were active at rates about the same as that of some of the best pyridine backbone compounds, (~ 10–50 g ha). In contrast, the other regioisomers, such as **128** or **129** were markedly less active. That structures with the backbone **128** gave poor results was consistent with the known SAR for all imidazolinones: any substitution ortho to the carboxylic acid adversely affects the activity. On the other hand, hetero[*b*]pyridines are known to be quite active compounds [27], so it was surprising that the corresponding heterobenzenes such as **129** were less herbicidal.

Further requirements are also evident. In the five-membered heterocycles fused to benzenes, it appears that at least two heteroatoms are required for the high activity. Even further, one of the heteroatoms must necessarily be located in the ortho position relative to the imidazolinone ring. Thus, the furan, **130**, and the pyrrole, **131**, were essentially inactive, and their isomers, **132** and **133**, were only slightly active. Nitrogen heterocycles with an unsubstituted NH, regardless of position, were inactive, and even the *N*-alkylated compounds, **135**, isomeric

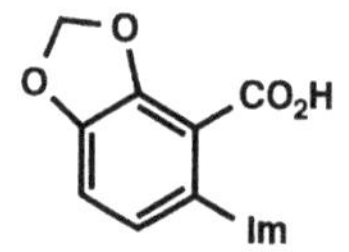

128

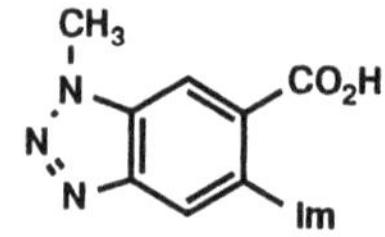

129

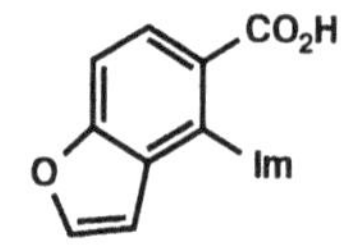

130

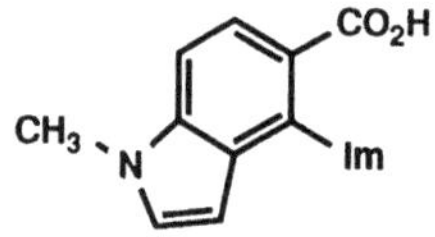

131

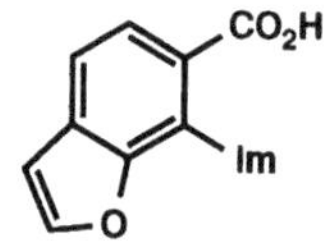

132

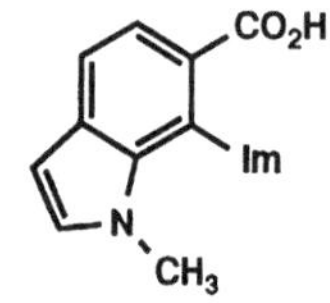

133

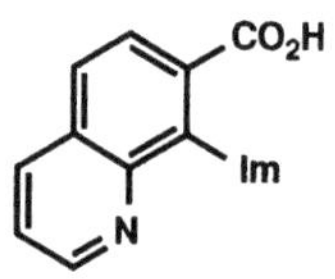

32

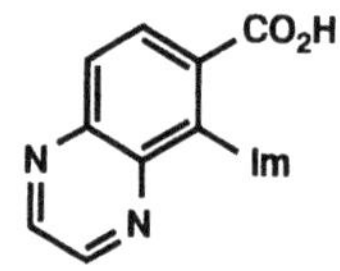

134

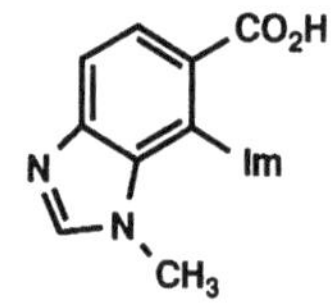

135

with **33** (R = Me), lost activity. The fused six-membered heterocyclic compounds were generally much less active than the five. The pyridine, **32**, had outstanding activity in some species, however. Actually this positional isomer of imazaquin is the only other one of ten to have useful herbicidal activity. The quinazoline **134** was also less active than **32**.

The inactivity of NH compounds has been previously observed for pyrrolopyridines [66] and other nitrogen containing compounds, and is likely to be due to metabolic deactivation. The sensitivity of the ortho substituent to the imidazolinone is probably a steric requirement for AHAS binding, but the requirement for an ortho heteroatom is more likely an electronic effect. This is similar to the differences seen for the various pyridine isomers and this phenomenon is discussed further in sect. 3.4.

3.1.4 SAR of Recent Novel 5- and 6-Substituted Imidazolinylpyridinecarboxylic Acids

Compounds in this area have been discussed at length and were the subject of a QSAR study correlating greenhouse weeds to physicochemical parameters [67]. More recently, a select set of 5-substituted compounds were described by an equation relating the activity in *Arabidopsis thaliana* to lipophilicity and the Hammett electronic parameter, σ_p [68]. The result, Eq. (1), was consistent with the earlier model by implying that highest activity in this series is with compounds with electron-donating 5-substituents, with a π value between 0.0 and 1.0. The 5-substituted pyridine analogs which have been prepared for a variety of reasons generally agree with the optimum for lipophilicity. There is less generality found for the electronic portion of the model.

$$\text{Log } 1/C \text{ arabidopsis} = 4.69 + 0.37\,\pi - 1.18\,\sigma_p - 0.26\,\pi^2 \quad (1)$$

$$R^2 = 0.89 \quad n = 18 \quad s = 0.24 \quad \text{F-test} = 38.42$$

Some qualitative observations can be made about some of the newer 5-substituted pyridine compounds. The 5-carboxylic acid derivative as well as the formyl, **71**, azido, **75**, carbamoyl, **83**, and amidino, **92**, compounds were inactive at a rate of 1 kg/ha. Somewhat more active were the hydrazones **79** and **91**, semicarbazones and simple imine derivatives. Fairly good activity was found for the esters, **68**, ortho esters, **78**, and acetals, **62** and **73**. The 5-formyl oximes, **69** and **70**, were particularly herbicidal, especially their *O*-alkylated derivatives [69], and a striking difference was observed between them and the 5-acetyl oximes [70]. Labile compounds, e.g., **62**, often showed more activity postemergence compared to pre-, presumably because of soil instability. Alkenyl and alkynyl analogs e.g., **63**, **67**, and **72**, were in most cases less active, although **64** retained high activity. Most of the heterocyclic substituents, e.g., **65**, **66**, **77**, and **80**, were quite good, with the exception of the inactive tetrazole **93**. The best of these was the 5-pyrazolyl analog, **76**, which was even more active than the 5-pyrrolyl compound, **85**.

The level of activity observed for the electron-withdrawing oximes and acetals was not anticipated by the previous QSAR, but the 5-carboxy, the 5-carbamoyl and the 5-acetyl compounds were more in line with prediction.

Finally, a compound in this series, the previously disclosed AC 263,222 is currently in development primarily because of its unusual effectiveness against sicklepod (*Cassia obtusifolia*) in leguminous crops.

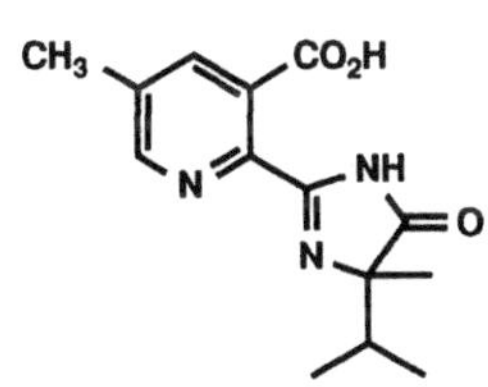

AC 263,222

3.2 Structure–Activity Relationships of the Sulfonylcarboxamides

The structure–activity effects seen in the sulfonylcarboxamide analogs were quite distinct from imidazolinones. As a class, they were lower in activity, although certain compounds were quite effective. The generic structure **136** indicates the different structural domains whose effects on activity were studied. The results are summarized below, but more detail can be found in a recent publication [60].

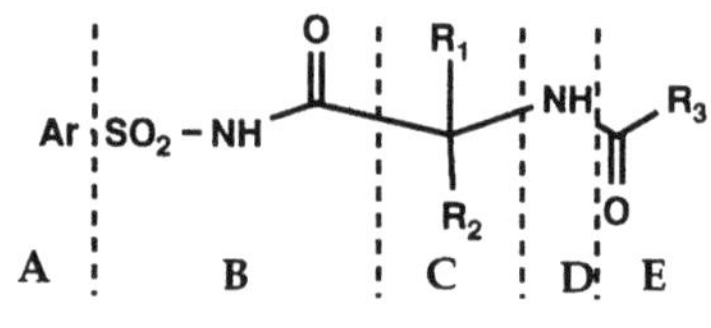

136

Region **A** was optimum with Ar = substituted phenyl. The best phenyl substituents were mono-, di- or trihalo, as well as alkyl. Methoxy was active only at the ortho position. Heterocyclic replacements for phenyl, or other replacements for Ar did not produce significant activity.

Replacement of the sulfonamido group **B** by carbonyl or phosphoryl gave inactive compounds. Hydrazides of type $-SO_2NHNHC(=O)$- and imidates of type –SO2–N=C(OR)– were likewise inactive.

In the **C** region, the R_1R_2 was optimal for [isopropyl, methyl], although [ethyl, methyl] was fairly active. In contrast to imidazolinones, where greatest activity was found for the **R** enantiomer, the **S** isomer of the sulfonylcarboxamides (when **E** = acetyl) was shown to be solely responsible for activity. This observation implies that the binding to the AHAS enzyme for the sulfonylcarboxamides is quite different from the imidazolinones.

A substituent on nitrogen in the **D** region gave less active compounds.

The **E** region in the initial discovery was an ortho carboxy pyridine compound, **30**, but it was the acetyl compounds which were more active and more heavily studied. Unsubstituted pyridines, benzenes or other heterocyclic amides other than acetyl had little or no activity. Groups which did have significant activity at this position were methoxycarbonyl and methylcarbamoyl. Longer chain esters or carbamates were less active, although they retained activity in the AHAS screen.

In summary, the most active sulfonylcarboxamides were **136**, where R_3 = methyl; R_1,R_2 = isopropyl, methyl; and where Ar = 4-chlorophenyl, 4-methylphenyl, 2,4-dichlorophenyl.

3.3 AHAS Inhibition

Evidence that the imidazolinones act by the uncompetitive inhibition of acetohydroxyacid synthase has been well described [1c] [7]. The recent work in this area has been aimed at determining the precise reasons for plant death. Previous work in sulfonylureas has suggested that it is the buildup of the substrate, α-ketobutyrate, which causes the toxic effect in plants. This is based on the apparent involvement of this metabolite in the bactericidal activity of sulfonyl ureas, although its in vivo plant activity was not previously known [71]. Shaner and others have recently shown that α-ketobutyrate is not responsible for plant death and conclude that plants die as a direct result of starvation for the branched-chain amino acids in the growing tissues [72]. This is supported by work which has shown that even when α-ketobutyrate production is inhibited by the action of isoleucine on threonine dehydratase, no corresponding decrease in the toxicity of imidazolinones was observed. The toxic effects are reversed only by the addition of all three branched-chain amino acids, leucine, isoleucine and valine [73]. Because the mature plant tissue is unaffected by AHAS inhibitors, the conclusion is that the new synthesis of these three amino acids does not occur in mature tissue. In the meristematic tissue, an AHAS inhibitors causes the disruption of protein synthesis and cell division within a few days of treatment. It has also been shown that AHAS inhibitors cause a decrease in translocation of photosynthate to the growing points, further depriving the plant of carbon. It is therefore likely that this combination of factors is responsible for phytotoxicity.

3.4 Uptake and Translocation Studies

While inhibition of AHAS varies somewhat among the various classes of herbicides known to disrupt this enzyme, it is clear that inhibition is a necessary but insufficient condition for herbicidal activity. To act at the whole plant level, a successful compound must be effectively absorbed and translocated to the growing points. Because of differences in their root and foliar absorption, imazapyr and imazaquin, for example, have different optimum application methods: imazaquin is readily taken up by roots facilitating its preemergence utility; imazapyr is rapidly absorbed by the foliage and is effective as a postemergence TVC.

Physicochemical properties of imidazolinones have been shown to be important for whole plant activity but they are poorly correlated with in vitro enzyme activity. For example, the relative order of the herbicidal activity of the compounds with benzene, quinoline and pyridine backbones is opposite to their relative enzyme activity. Uptake, transport and metabolism must therefore be significantly dependent on some of these physicochemical properties. It is known that imidazolinones with an unmasked carboxylic acid utilize the ion-trapping (or weak acid) mechanism [74] for uptake and transport. Indeed, uptake experiments with imidazolinones in dinitrophenol-ruptured cells have shown that absorption by passive transport is not significant [75]. More importantly, a detailed study has now been done to determine the relationship between physicochemical properties on uptake and translocation into corn and sunflowers, and these will be reviewed in sects. 3.4.1 and 3.4.2 [76].

Sulfonylcarboxamides are more poorly absorbed than imidazolinones, and thus behave more in this regard like sulfonylureas. No definitive study has been conducted, but an obvious reason for this behavior is their highly crystalline nature and low solubility.

3.4.1 Uptake Studies of Imidazolinones

A study of compound movement in hydroponically grown sunflower and corn plants was undertaken, selecting a set of twelve 5-substituted imidazolinones whose physicochemical properties covered a wide range of data space and were not co-correlated with each other [75]. The results of the uptake study in both species were equations which point to a strong dependence on lipophilicity for uptake. The relationships are fairly linear, except for very highly lipophilic compounds:

$$\text{Log } 1/C_{\text{corn uptake}} = 1.39 + 0.63\pi - 0.17\pi^2 \quad (2)$$

$$R^2 = 0.87 \quad n = 12 \quad s = 0.28 \quad \text{F-test} = 31.03$$

$$\text{Log } 1/C_{\text{sunflower uptake}} = 1.47 + 0.46\pi - 0.10\pi^2 \quad (3)$$

$$R^2 = 0.87 \quad n = 12 \quad s = 0.21 \quad \text{F-test} = 30.17$$

No dependence on steric or electronic parameters was found in either species. A similar correlation could be obtained with 16 unlabeled compounds by using the disappearance from the hydroponic solution as the dependent variable.

Uptake was also measured for the regioisomeric compounds of imazapyr, imazaquin, imazethapyr and AC 263,222. These regioisomers (**137**–**140**) are all AHAS inhibitors but are only slightly active in whole plants and are only slightly absorbed by roots.

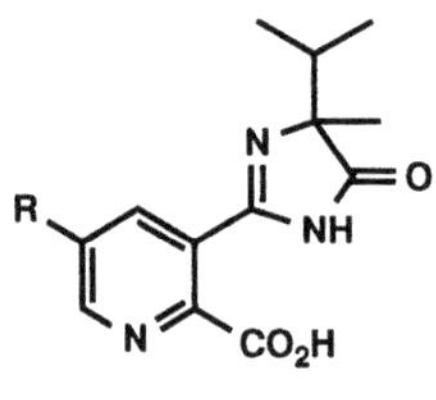

137 R = H
139 R = Me
140 R = Et

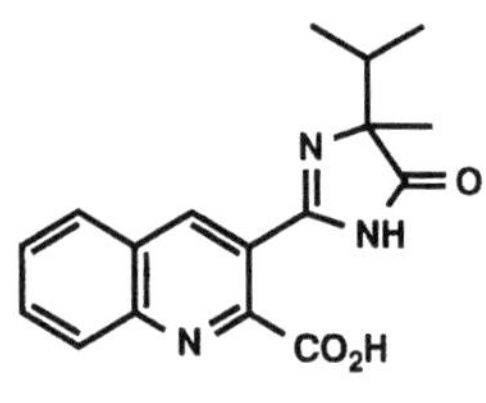

138

These measurements confirm that the poor in vivo activity of these isomers is due to very poor uptake. It is unclear which of the physicochemical properties is responsible: each of the isomers have similar values for LogP and pK_a. Infrared spectroscopic evidence, however, suggests that **137–140** form strongly zwitterionic species, so one explanation is that the concentration of the neutral (free acid) species for the isomeric compounds is much lower than for the herbicidal isomers. It is well known, for example, that a zwitterion will have a LogP value as much as 4.0 units lower than the neutral species [77]. In addition, other experiments show that **137–140** form metal complexes more strongly than the active compounds [78].

In any event, soil pH was recently shown to be a factor for root absorption of AC 263,222 [79], so it is clear that the nature of the species is important, either by permitting release from the soil or for insuring uptake by having the correct LogP.

3.4.2 Translocation Studies of Imidazolinones

Translocation from roots to shoots in sunflowers was also measured for a set of 5-substituted pyridine imidazolinones [75]. In addition to lipophilicity, an electronic parameter, F, was also a significant factor, and the equation below was one of the better models obtained:

$$\text{Log}\ (\%\ \text{Translocation})_{\text{sunflower}} = 1.33 + 0.20\,\pi + 0.95\,\text{F} \quad (4)$$

$$R^2 = 0.85 \qquad n = 11 \qquad s = 0.08 \qquad \text{F-test} = 22.70$$

A similar study in corn was unsuccessful because of the rapid metabolism of several of the labeled compounds.

It was expected that pK_a, influenced by the electronic nature of the substitution, is also an important factor for transport. Although pK_a was measured for these compounds, it was not well-correlated with either transport or F for these eleven compounds. This translocation model was tested for 5-bromopyridine imidazolinone, a compound both lipophilic and high in electronic F value. Subsequent measurement of this compound showed much poorer translocation than predicted. Apparently there are other factors important for translocation (e.g., steric) which could not be included in this 11-compound model.

Uptake and translocation has also been shown to be affected by admixture of other herbicides or surfactants as discussed at length in [80].

3.5 Metabolism of Imidazolinones

In addition to importance for environmental studies and registration requirements, metabolism studies in imidazolinones have continued in order to understand crop selectivity and the mechanism of herbicide safeners.

3.5.1 Metabolism and Modes of Selectivity

Selectivity of imidazolinones has been exclusively attributed to the metabolism of the compounds by resistant species. The metabolic deactivation of imazethapyr has been shown to be benzylic hydroxylation followed by glycosylation and incorporation into cell walls [9]. As might be expected, imazapyr, which lacks an alkyl side chain, is a non-selective compound. The exact behavior of imazaquin in soybean, on the other hand, has been studied more recently [81]. The major metabolite **143** was isolated, along the lesser amounts of **141** and **142**, and a minor amount of **144**. The assumption is that an intermediate, e.g., **145**, is formed which undergoes ring opening to **141** or reduction to **142**. Further ring opening of **141** leads to the two isomers, **143** and **144**. The difference in metabolism between imazaquin and imazethapyr has been ascribed not only to their structure, but to the difference in lipophilicity of the herbicides which may influence their availability to different detoxifying enzymes in the plant [81].

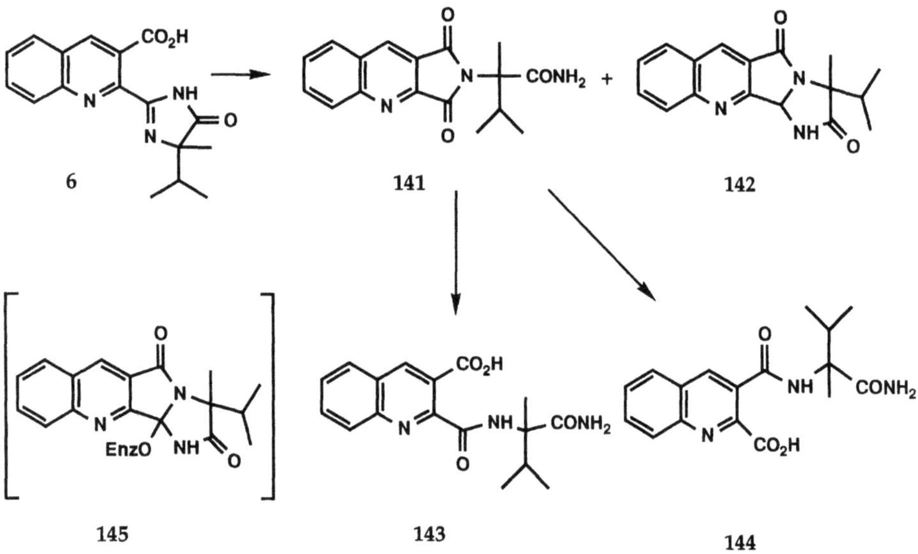

3.5.2 Naphthalic Anhydride as a Safener for Imidazolinones

The metabolism of imazethapyr has been shown to be affected by naphthalic anhydride or its derivatives [82]. Although corn rapidly converts imazethapyr to **146** which is further glycosylated to **147**, the process is not sufficiently rapid to make **5** useful in corn hybrids which have not been specifically selected for tolerance. The study determined that naphthalic anhydride and its derivatives accelerate this hydroxylation step and selectivity in corn is then observed.

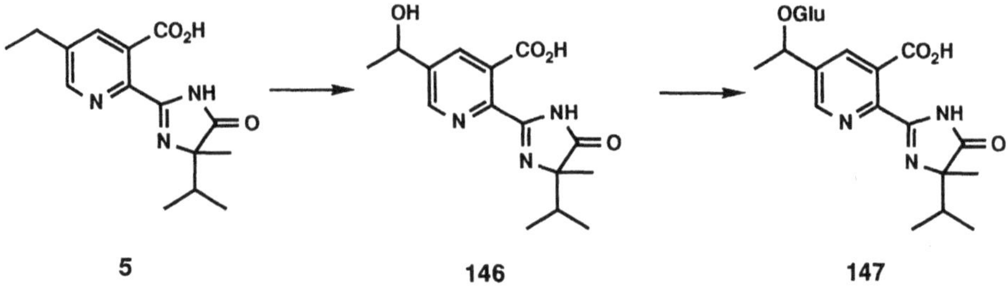

Other compounds have been claimed to have similar safening activity in imidazolinone and other herbicides [83]. It appears that these generally work by inducing mixed function oxidases in certain plants.

3.6 Weed and Crop Resistance to Imidazolinones

The importance of this class of herbicides has led to their widespread use in certain crops, particularly soybeans. Biotechnology has permitted further expansion of their utility through the development of herbicide-resistant hybrids. Resistance in certain weeds has also been observed for these and other AHAS-inhibiting compounds.

3.6.1 Imidazolinone-Tolerant Crops

Although imazamethabenz methyl is used in cereals, and imazaquin and imazethapyr have been remarkably successful in soybeans, the imidazolinones have only recently been developed in other major crops. This advance was achieved through the development of resistant crop hybrids rather than the traditional analog synthesis approach.

Two successful programs have been developed in corn: one by Cyanamid in collaboration with Pioneer Hybrids [84] and one by ICI/Garst Seed [85]. These corn hybrids have recently been introduced for use with imazethapyr. Similarly, a seed mutagenesis program has recently been reported in which an imidazolinone tolerant wheat mutant has been identified [86]. This has been successfully field tested using imazethapyr, **5**.

The mechanism of selectivity in all cases is due to a single gene transformation, resulting in a change in the nature of the AHAS enzyme in the resistance lines. These and similar studies with lines tolerant to sulfonylureas have shown that there is sometimes, but not always, cross-tolerance among the AHAS-inhibiting herbicides. This implies that some, but not all, of the herbicide binding regions of the AHAS protein are shared for the two different classes of compounds. The vigor of the imidazolinone-tolerant plants also reconfirms the previous finding that the binding of the herbicide is not at the catalytic site, because enzyme function is unaffected.

3.6.2 Imidazolinone-Resistant Weeds

As the use of imidazolinones and other AHAS-inhibiting herbicides has increased in existing and new markets, the issue of herbicide resistance has become increasingly important. The occurrence of resistance in weed populations subjected to selection pressure of these compounds has been reported [87]. Even in the absence of a resistant biotype, the use of a single compound can shift the spectrum of weeds from susceptible ones to one of mostly tolerant weeds. To address these issues, a multi-company working group has been established, and has published guidelines for managing resistance [88, 89].

In their report, the group summarizes the contributing factors leading to field resistance. These include the loss of older products; the use of a single herbicide rather than mixtures; decreased cultivation as a weed control measure; and sole reliance on herbicide products with the same mode of action. Some of these factors can be dealt with by judicious agricultural techniques, such as practicing crop as well as herbicide rotation, avoiding seed contamination and careful monitoring of the weed populations. A program of management employing these recommendations should prevent the premature loss of these very important herbicide classes.

3.7 Comparison of the Behavior of Imidazolinone and Sulfonylcarboxamides to other AHAS Compounds

Similarity in mode of action among imidazolinones, sulfonylureas, pyrimidyl aryl ethers and triazolopyrimidine sulfonamides has not meant similarity in end use, selectivity or use rate. The chief reasons are different de-toxification mechanisms available to each class of compounds. In addition, the translocation of each class can be substantially different, and it is clear that they do not all bind to the enzyme in exactly the same fashion. Developing molecular models for the shape of the binding site using the active analogs has been a formidable task, and the complexity of the enzyme itself has not yet allowed a determination of the tertiary structure. Judging by the number and content of issued patents, imidazolinones continue to be more restricted than the sulfonylureas in terms of the number of highly active structural types. The pyrimidyl aryl ethers, triazolopyrimidine sulfonamides and even sulfonylcarboxamides also appear to tolerate greater structural diversity.

4 Outlook for Imidazolinones and Related Compounds

The imidazolinone class of compounds now consists of four commercial products, with a fifth compound AC 263,222 in development. Sulfonylcarboxamides have not yet provided a developmental candidate.

Improvement in these classes of chemistry is still possible, with work focusing on the search for analogs with improved selectivity, environmental properties, or effectiveness on problem weeds. It is likely that further improvements in the routes to the commercial compounds will continue especially if others are introduced. A new series of highly active compounds based on imidazolinones could yet be discovered, but experience has taught that herbicidal activity falls off rapidly with drastic structural changes. Genetic manipulation will therefore continue as an important means to expand the utility of the compounds in a wider variety of crops.

The exact nature of imidazolinone binding and inhibition to the enzyme is still a worthwhile area for study. A detailed understanding of this binding site may be required before further improvements in synthetic analogs can be made. New imidazolinone analogs will also be faced with the problem of resistance; the impact of each new compound on the existing products will have to receive serious attention.

Acknowledgement. The author thanks his many colleagues at American Cyanamid whose research is reviewed above and gratefully acknowledges Gerald Berkelhammer, Michael Condon, Robert Doehner, Marinus Los and Dale Shaner for their editorial assistance of the manuscript.

5 References

1. (a) Wepplo P (1990) Pest 29: 293; (b) Ladner DW (1990) Pest Sci 29: 317; (c) Stidham MA, Shaner DL (1990) Pest Sci 29: 335
2. Shaner DL, O'Connor SL (eds) (1991) The imidazolinone herbicides. CRC, Boca Raton Ann Arbor Boston London
3. Los M, Kust CA, Lamb G, Diehl RE (1980) Hortic Sci 15: 22
4. Los M (1984) In: Magee PS, Kohn GK, Menn JJ (eds) Pesticide synthesis through rational approaches, ACS Symp. Ser. No. 255, American Chemical Society, Washington, pp 29–44
5. Los M (1989) US 4 798 619
6. Reference 2, chaps 17–20, pp 211–256
7. Reference 2, chap 6, pp 53–70
8. Stidham MA (1991) in: Baker NR, Percival MP (eds) Herbicides. Elsevier Science Publishers, New York, pp 247–266
9. Reference 2, chap 11, pp 161–165
10. Reference 2, chap 14, pp 179–182
11. Singh BK, Flores HE, Shannon JC (eds) (1992) Biosynthesis and molecular regulation of amino acids in plants. American Society of Plant Physiologists, Rockville, Maryland (Current topics in plant physiology, vol 7)
12. Bayer (1988) DE 3 633 952
13. Hoffmann-La Roche (1986) EP 183 993
14. Ciba-Geigy (1987) EP 213 079 A
15. Ciba-Geigy (1987) EP 215 738 A
16. Mitsubishi Chemical Industries (1988) J63 079 878 A
17. Mitsubishi Chemical Industries (1988) J63 303 972 A
18. Mitsubishi Chemical Industries (1988) J61 171 471 A
19. Mitsubishi Chemical Industries (1986) J64 003 183 A
20. Schering (1986) DE 3 520 260 A
21. Bayer (1988) DE 3 634 887 A
22. Ciba-Geigy (1988) DE 3 828 541 A
23. Takenaka H, Hayase Y (1989) Heterocycles 29: 1185

24. Kehne H (1990) in: Frehse H, Hesseler-Schmitz E, Conway S (eds) Seventh international congress of pesticide chemistry, Hamburg, 5–10 Aug 1990, Abstract 01B-16, p 108
25. Ciba-Geigy (1989) EP 298 029 A
26. Guaciaro MA, Los M, Little DL, Marc PA, Quakenbush L (1992) in: Baker DR, Fenyes JG, Steffens JJ (eds) Synthesis and chemistry of agrochemicals III. ACS Symposium Series, 504 American Chemical Society, Washington, DC, pp 56–74
27. American Cyanamid (1987) EP 227 932 A
28. Nissan Chemical Industries (1987) US 4 696 694
29. American Cyanamid (1992) US 5 108 485
30. American Cyanamid (1992) ZA 9 102 722
31. American Cyanamid (1993) US 5 180 419
32. Ciba-Geigy (1987) EP 233 150 A
33. Nissan Chemical Industries (1989) US 4 824 474
34. Ciba-Geigy (1989) EP 327 492 A
35. Mitsui Toatsu (1990) J02 202 886 A
36. Bayer (1992) EP 517 052 A
37. (a) Celamerck (1988) 4,723 989 (b) Sumitomo Chemical Industries (1987) AU 8 661 073 A (c) Idemitsu Kosan (1988) US 4 726 835
38. Otsuka Kagaku Yakuhin (1987) J6 1183-272 A
39. Schering (1990) DE 3 913 757 A
40. Stauffer (1986) US 4 576 631 A
41. American Cyanamid (1987) EP 254 951 A
42. Schering (1987) EP 216 360 A
43. Ciba-Geigy (1988) EP 296 109 A
44. Nissan Chemical Industries (1987) J62 174 069 A
45. Cossette M (1992) Joint Middle Atlantic Regional and Southestern Regional Meeting, Alexandria VA, 7 Dec 1992
46. Doehner R (1991) Pestic Sci 31: 581
47. Hoechst Celanese (1991) EP 456 504 A
48. ICI (1989) EP 308 084 A
49. Shell (1988) EP 292 032 A
50. Ciba-Geigy (1988) US 4 748 244
51. Wacker Chemie (1990) EP 371 499 A
52. Hoechst Celanese (1991) EP 425 767 A
53. (a) Kuraray (1991) J03 101 660 A (b) Kuraray (1991) J03 101 674 A
54. Davies LB, Leci OA, Sammes PG, Watt RA (1978) J Chem Soc Perkin Trans 1, 1293
55. Neunhoeffer H, Metz HJ (1983) Liebigs Ann Chem 1476
56. Seitz G, Görge L, Dietrich S (1985) Tetrahedron Lett 26: 4355
57. (a) Taylor EC, Macor JE (1985) Tetrahedron Lett 26: 2419 (b) Taylor EC, Macor JE (1986) Tetrahedron Lett 27: 431
58. Brien DJ, Naiman A, Vollhardt KPC (1982) J Chem Soc 133
59. Grigg R, Scott R, Stevenson P (1982) Tetrahedron Lett 23: 2691
60. Alvarado SI, Crews AD, Wepplo P, Doehner EF, Brady TE, Gange DM, Little DL (1992) in: Baker DR, Fenyes JG, Steffens JJ (eds) Synthesis and Chemistry of Agrochemicals III, ACS Symposium Series 504, American Chemical Society, Washington, chap 8, pp 75–80
61. American Cyanamid (1989) US 4 883 914
62. American Cyanamid (1991) US 4 992 094
63. Kumiai (1989) J01 250 343 A
64. Kumiai (1989) J01 224 355 A
65. Cloudsdale, IS1, Anderson RJ, Chinn GW, Deng PN, Herberich-Patton JC, Pomes JC (1993) 205th ACS National Meeting, Denver, AGRO Abstract 33

66. Finn J, Quinn, N, Buckman B (1991) in: Baker DR, Fenyes JG, Moberg WK (eds) Synthesis and chemistry of agrochemicals II. ACS Symposium Series 443, American Chemical Society, Washington, DC, chap 11, pp 133–143
67. Cross B, Ladner DW (1992) in: Draber W, Fujita T (eds) Rational Approaches to Structure, Activity, and Exotoxicology of Agrochemicals CRC Press, Boca Raton FL, chap. 13, pp 331–355
68. Little DL, Ladner DW, Shaner DL, Ilnicki RD, Pestic Sci, in press
69. American Cyanamid (1993) US 5 215 568
70. Nissan Chemical Industries (1987) J62 174 069
71. LaRossa RA, VanDyk TK, Smulski DR (1990) in: Barak Z, Chipman DM, Schloss JV (eds) Biosynthesis of branched-chain amino acids, Balaban Publishers, New York, pp 109–121
72. Shaner DL (1992) in: Singh BK, Flores HE, Sahnnon JC (eds), Biosynthesis and molecular regulation of amino acids in plants, American Society of Plant Physiologists, Rockville, Maryland (Current topics in plant physiology, vol 7) pp 174–183
73. Singh BK, Shaner DL (1992) ibid. Singh BK, Flores HE, Shannon JC (eds), Biosynthesis and molecular regulation of amino acids in plants, American Society of Plant Physiologists, Rockville, Maryland (Current topics in plant physiology, vol 7) pp 354–357
74. Van Ellis MR, Shaner DL (1988) Pestic Sci 23: 25
75. Little DL (1991) Physico-chemical parameters affecting root absorption and translocation of 5′-substituted pyridine imidazolinones, Thesis, Rutgers University, New Brunswick NJ
76. Little DL, Shaner DL, Ladner DW, Tecle B, Ilnicki RD (1990) in: Frehse H, Kesseler-Schmitz E, Conway S (eds) Book of Abstracts, Seventh International Congress of Pesticide Chemistry, Hamburg, 5–10 Aug 1990, Abstract 1C-21, p 161
77. Hansch C, Leo A (1979) Substituent constants for correlation analysis in chemistry and biology. Wiley-Interscience, New York
78. Rajamoorthi K, Shaner DL, Rajan S, Stockton GW (1993) private communication
79. Tuxhorn G, Deane J, Shaner DL (1993) WSSA Meeting Abstracts, Denver, 8–11 Feb, 1993, Vol 33, Abstract 343
80. Shaner DL (1989) Factors affecting soil and foliar bioavailability of imidazolinones. American Cyanamid Company, Princeton, NJ
81. Tecle B, DaCunha A, Shaner DL (1993) Pestic Biochem and Physiol in press
82. Shaner DL (1991) Z Naturforsch 46c: 893
83. Ciba-Geigy (1985) EP 159 287 A
84. Newhouse K, Singh B, Shaner D (1991) Theor Appl Genet 83: 65
85. Bright SWJ (1992) in: Singh BK, Flores HE, Shannon JC (eds) Biosynthesis and molecular regulation of amino acids in plants, American Society of Plant Physiologists, Rockville, Maryland (Current topics in plant physiology, vol 7) pp 184–194
86. Newhouse KE, Smith WA, Starrett MA, Schaefer TJ, Singh BK (1992) Plant Physiol 100: 882
87. Devine MD, Hall LM (1990) in: Frehse H, Kesseler-Schmitz E, Conway S (eds) Book of Abstracts, 7th International Congress of Pesticide Chemistry, 5–10 Aug 1990. Hamburg, 4D-03, p 385
88. ALS/AHAS Inhibitor resistance working group (1993) Guidelines for managing resistance to ALS/AHAS-inhibiting herbicides (technical bulletin available from American Cyanamid, Ciba-Geigy, Dow-Elanco, DuPont, Hoechst, Ishihara Chemical, Kumiai Chemical, Nissan Chemical Industries, Monsanto, and BASF)
89. Shaner DL (1992) in: Copping LE, Green MB, Rees RT, Pest Management in Soybeans, Elsevier Applied Science, London, pp 339–347

Triazolopyrimidine Sulfonanilide Herbicides and Related Compounds

William A. Kleschick

DowElanco Research Laboratories, 306/E2, P.O. Box 68955,
9410 Zionsville Road, Indianapolis, Indiana 46268-1053, USA

The synthesis, chemistry and structure–activity relationships in the triazolopyrimidine sulfonanilide class of acetolactate synthase inhibiting herbicides is reviewed. Sulfonanilides based on the 1,2,4-triazolo[1,5-*a*]pyrimidine, 1,2,4-triazolo[1,5-*c*]pyrimidine, imidazolo[1,2-*a*]pyrimidine, pyrazolo[1,5-*a*]pyrimidine, 1,2,4-triazolo[1,5-*a*]-1,3,5-triazine, pyrazolo[1,5-*a*]-1,3,5-triazine and thiazolo[3,2-*b*]-1,2,4-triazole ring systems are included in this review. Aryl sulfonamide derivatives of 2-amino-1,2,4-triazolo[1,5-*a*]pyrimidine, 2-amino-1,2,4-triazolo[1,5-*a*]-1,3,5-triazine and 2-amino-1,2,4-pyrazolo-[1,5-*a*]pyrimidine are also covered.

Chemistry of Plant Protection, Vol. 10

List of Symbols and Abbreviations

Ac	Acetyl
ALS	Acetolactate synthase
Ar	Aryl
Bn	Benzyl
t-Bu	*tert*-Butyl
DMF	*N*,*N*-Dimethylformamide
DMSO	Dimethylsulfoxide
Et	Ethyl
Me	Methyl
NBS	*N*-Bromosuccinimide
NCS	*N*-Chlorosuccinimide
OM	Organic matter
Ph	Phenyl
i-Pr	*i*-Propyl
n-Pr	*n*-Propyl
THF	Tetrahydrofuran
TMS	Trimethylsilyl

1 Introduction

The 1,2,4-triazolo[1,5-*a*]pyrimidine sulfonanilide class of herbicides is the product of a research effort begun in 1982 and initially aimed at preparing bioisosteric analogs of sulfonylureas [1]. Members of this class of compounds include flumetsulam (DE-498, **1**) and DE-511 (**2**) [2, 3]. The ISO proposed common name for DE-511 is metosulam. Flumetsulam is under development for broadleaf weed control in corn and soybeans and metosulam is being developed for broadleaf weed control in cereals and maize.

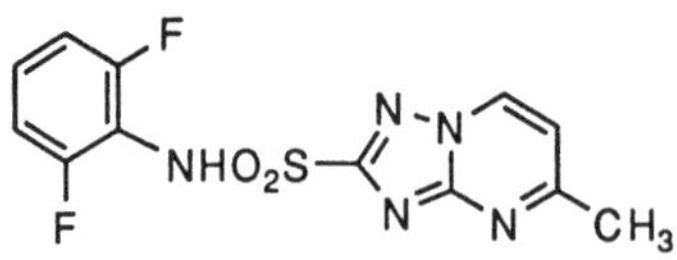

1

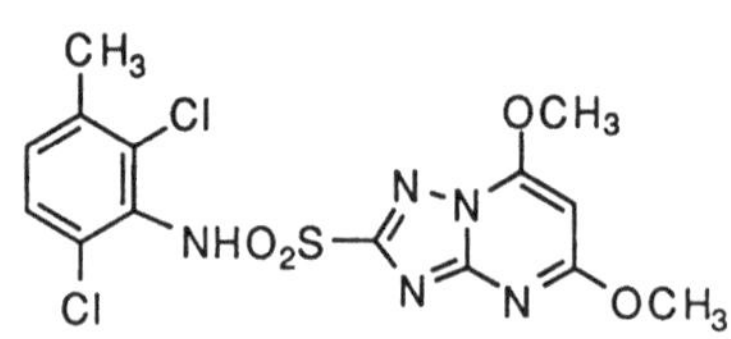

2

Since the discovery of the 1,2,4-triazolo[1,5-*a*]pyrimidine sulfonanilides, the scope of the class has increased to include sulfonanilide derivatives of other bicyclic heterocyclic ring systems containing bridgehead nitrogen atoms.

These ring systems include 1,2,4-triazolo[1,5-*c*]pyrimidine, imidazolo[1,2-*a*]-pyrimidine, pyrazolo[1,5-*a*]pyrimidine, 1,2,4-triazolo[1,5-*a*]-1,3,5-triazine and thiazolo[3,2-*b*]-1,2,4-triazole. This review outlines the synthesis, chemistry and structure–activity relationships associated with this growing class of herbicidal inhibitors of acetolactate synthase (ALS, EC 4.1.3.18) [4]. Discussions of structure-activity relationships cover in vitro activity against ALS and soil decomposition data in addition to herbicidal activity.

2 Synthesis and Chemistry

2.1 (*N*-1,2,4-Triazolo[1,5-*a*]pyrimidine)benzene Sulfonamides

The first derivatives of 1,2,4-triazolo[1,5-*a*]pyrimidine which exhibited significant levels of herbicidal activity were aryl sulfonamide derivatives of the 2-amino compound (**3**) [1, 5]. The synthesis of these compounds is accomplished by one of two synthetic routes outlined in Scheme 1 [5–10]. The route beginning

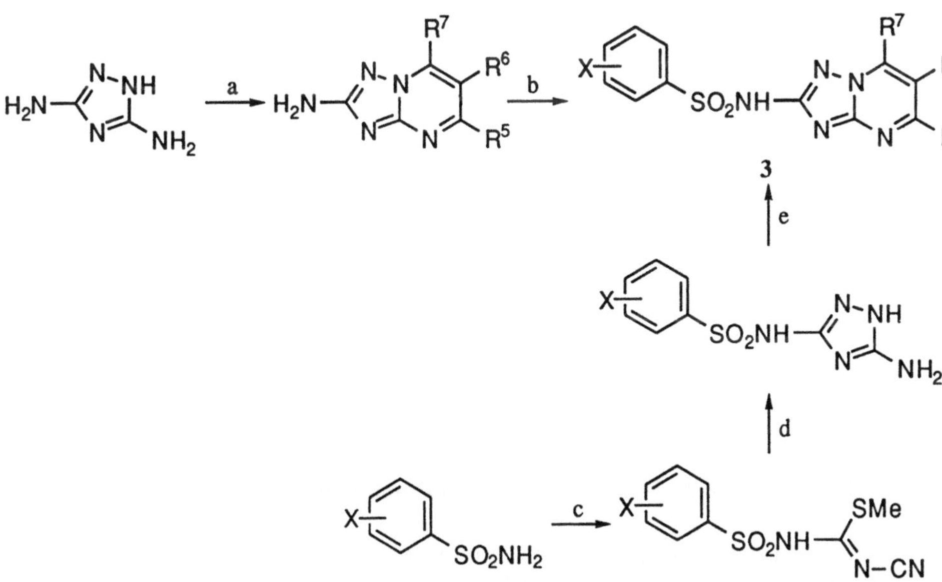

Scheme 1. (a) $R^5COCHR^6COR^7$; (b) arylsulfonyl chloride, pyridine; (c) base, dimethyl *N*-cyanodithioiminocarbonate; (d) N_2H_4; (e) $R^5COCHR^6COR^7$

with 3,5-diamino-1,2,4-triazole is preferred due to simplicity where X is a hydrogen atom or electron-donating group and A, B and C are not highly sensitive or reactive functional groups. The highly insoluble nature of 2-amino-1,2,4-triazolo[1,5-*a*]pyrimidines makes the harsh reaction conditions for the final step necessary. These harsh conditions are not compatible with the thermal stability of many benzenesulfonyl chlorides containing electron-withdrawing groups. The route starting from the substituted benzene sulfonamide is preferred due to higher yields in instances where X is an electron-withdrawing group and A, B and C are sensitive functional groups. The route is applicable to synthesis of the vast majority of compounds represented by structure **3**. Good to excellent levels of regioselection are achieved in condensations of the aminotriazole intermediates with unsymmetrical β-dicarbonyl compounds (i.e. $R^5 \neq R^7$, see below).

The related aryl sulfonamide derivatives of 2-amino-1,2,4-triazolo[1,5-*a*]-1,3,5-triazine (**4**) and 2-aminopyrazolo[1,5-*a*]pyrimidine (**5**) have also been prepared and reported to be active as herbicides [11, 12]. In addition related structures **6** and **7** have been reported to exhibit herbicidal activity, but doubt exists as to whether these compounds are herbicidally active or are pro-herbicides of sulfonylureas [13–16].

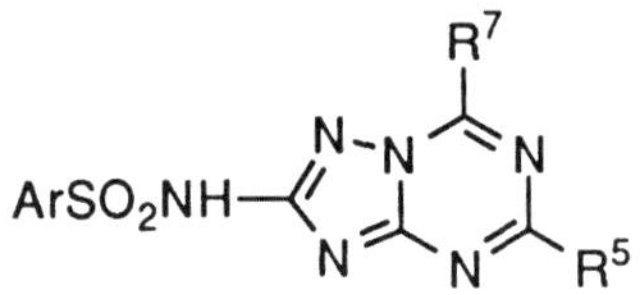

4

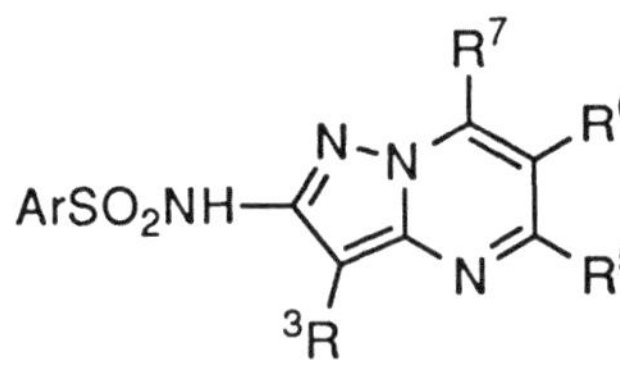

5

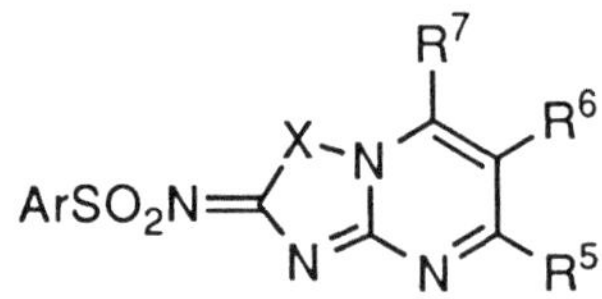

6: X = O
7: X = S

2.2 1,2,4-Triazolo[1,5-*a*]pyrimidine-2-sulfonanilides

The 1,2,4-triazolo[1,5-*a*]pyrimidine-2-sulfonanilides (**8**) test a hypothesis regarding the contribution of the $-SO_2NH-$ functionality in determining activity in compounds of structure **3** [17–27]. The general synthetic route to compounds from this class is illustrated in Scheme 2 [1, 28]. This route enables the preparation of a variety of compounds including those substituted with alkyl, aryl or haloalkyl groups (R^5, R^6 and R^7) on the triazolopyrimidine ring. It is also applicable to the synthesis of compounds containing a wide variety of functional groups in the 6-position (R^6) other than alkyl, aryl or haloalkyl groups. These substituents in the 6-position (e.g. halogen or alkoxy) are present in the starting

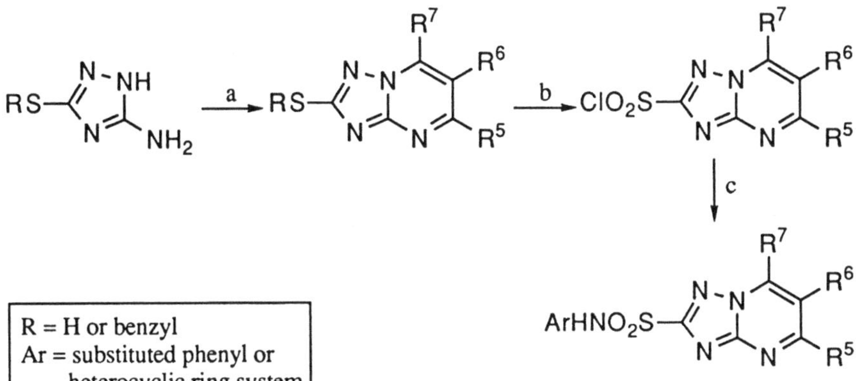

R = H or benzyl
Ar = substituted phenyl or heterocyclic ring system

8

Scheme 2. (a) $R^5COCHR^6COR^7$; (b) Cl_2, H_3O^+; (c) $ArNH_2$ or ArNHTMS, base

β-dicarbonyl compound or introduced by electrophilic substitution at the end of the synthesis (e.g. the conversion of **8**: R^5 = Me, $R^6 = R^7$ = H to **8**: R^5 = Me, R^6 = Br, R^7 = H with bromine in glacial acetic acid). The scope of β-dicarbonyl compounds operable in the synthesis include β-ketoaldehydes and β-diketones and a variety of equivalents such as malonaldehyde bisacetals, β-ketoacetals, methaniminium salts, 2,3-dihalocarbonyl compounds, mucochloric and mucobromic acid and other compounds commonly employed in the synthesis of pyrimidine rings.

When the aminotriazole intermediate is reacted with unsymmetrical β-ketoaldehydes and β-diketones, modest to very high levels of regioselection are observed [29–32]. In some instances (e.g. reaction with acetylacetaldehyde dimethyl acetal and other ketoaldehyde derivatives) the regiochemical outcome is determined by careful choice of reaction conditions (e.g. acid or base catalysis) to achieve good to excellent selectivity in either direction (e.g. ratios of 85:15 to 0:100) [29].

With a minor modification the synthetic route illustrated in Scheme 2 can be used to prepare compounds containing other functional groups in 5-, 6- and 7-positions [33]. Condensation of the aminotriazole intermediate with malonate ester derivatives or β-ketoester derivatives produces triazolopyrimidine rings substituted with hydroxyl groups (i.e. $R^5 = R^7$ = OH or R^7 = OH respectively). Condensation with β-ketoesters is regiospecific and yields the 7-hydroxy compound exclusively. The hydroxyl groups in these compounds are converted to chlorine by reaction with phosphorus oxychloride, and the chlorine atoms displaced with a variety of nucleophiles such as alkali metal alkoxides. The displacement of the chlorine atoms is possible either before or after conversion of the mercapto or benzylthio compounds to the sulfonyl chloride and reaction with the aryl amino compound to form the sulfonamide bond. Displacement of the chlorine atoms last is advantageous in instances where R^5 and/or R^7 are sensitive to the conditions for conversion of the mercapto or benzylthio compounds to the sulfonyl chloride (e.g. amino or alkylthio) or reaction of the sulfonyl chloride with the aryl amine (e.g. amino). In some cases *N*-silyl anilines are used for the final reaction with superior results [34].

The high degree of regioselection observed in the reaction of 3-amino-5-benzylthio-1,2,4-triazole with β-ketoesters to form 7-hydroxy compounds is useful in accessing specific substitution patterns on the triazolopyrimidine ring. Substitution patterns which are not obtained exclusively by direct condensation with unsymmetrical β-diketones or β-ketoaldehyde derivatives are obtained through this method (Scheme 3) [17–27, 30, 35]. For example, the 7-chloro compound derived from the initial condensation product reacts with Grignard reagents and reducing agents to afford intermediates substituted in the 7-position with alkyl or hydrogen atoms respectively.

In addition to the reduction of the halogens in the 7-position in the presence of relatively unreactive functional groups in the 5- and 6-positions, certain highly chemoselective reductions are possible (Scheme 4) [17–21, 23]. A chlorine atom in the 7-position is removed by reduction with zinc–copper couple in the presence of chlorine atoms in either the 5- or 6-positions. Selective reduction of the chlorine at the 7-position in presence of the 5-chlorine is particularly useful since the remaining 5-chlorine can be further reacted with nucleophiles to introduce new functionality.

The synthesis of triazolopyrimidine sulfonanilides (**8**) is accomplished in many circumstances by a route in which the triazolopyrimidine ring is formed at a late stage after formation of the sulfonamide bond (Scheme 5). The key intermediate in this route is the aminotriazole sulfonanilide **9**. This intermediate is obtained by one of three methods: 1) the oxidative degradation of the triazolopyrimidine ring; 2) an amino protecting group strategy for conversion of 3-amino-5-mercapto-1,2,4-triazole (**10**) to **9**; and 3) a direct conversion of **10** to **9** [36–38]. The latter method is the most efficient, with yields of 98% achieved in the conversion of **10** to **9** where Ar = 2,6-difluorophenyl. This approach is

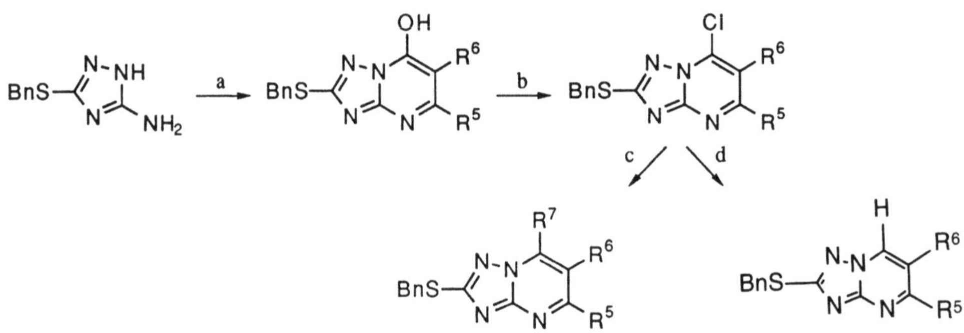

Scheme 3. (a) $R^5COCH(R^6)COOR$; (b) $POCl_3$; (c) R^7MgX; (d) Zn-Cu, AcOH (2 equiv.), MeOH-THF

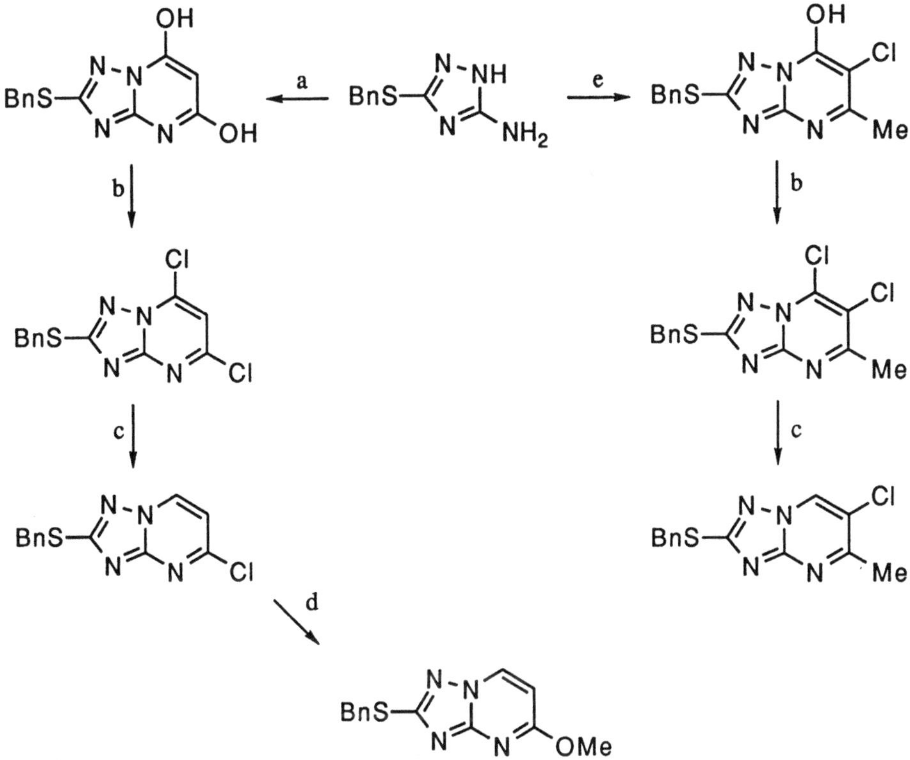

Scheme 4. (a) $CH_2(COOMe)_2$, NaOMe, MeOH-EtOH, reflux; (b) $POCl_3$, MeCN, reflux; (c) Zn-Cu, AcOH (2 equiv.), MeOH-THF; (d) NaOMe, MeOH; (e) MeCOCH(Cl)COOEt, AcOH, 100 °C

attractive in many situations where sensitive functional groups are present on the triazolopyrimidine ring. The regioselection in the condensation of intermediate **9** with a variety of unsymmetrical β-dicarbonyl compounds follows the same patterns established for the condensation with 3-alkylthio-5-amino-1,2,4-triazoles [39].

Another interesting example of selectivity in the functionalization of the triazolopyrimidine is seen in a variation in the basic route illustrated in Scheme 3 (Scheme 6). The 3-amino-1,2,4-triazole-5-sulfonanilide intermediate is condensed with dimethyl malonate and the condensation product is reacted with phosphorus oxychloride to yield the 5,7-dichloro substituted triazolopyrimidine. The 5,7-dichloro intermediate is then reacted in sequence with two equivalents of sodium methoxide followed by one equivalent of sodium methylmercaptide to yield the 5-methoxy-7-methylthio analog, or two equivalents of sodium methylmercaptide followed by one equivalent of sodium methoxide to yield the 7-methoxy-5-methylthio analog. In the latter transformation, benzyl

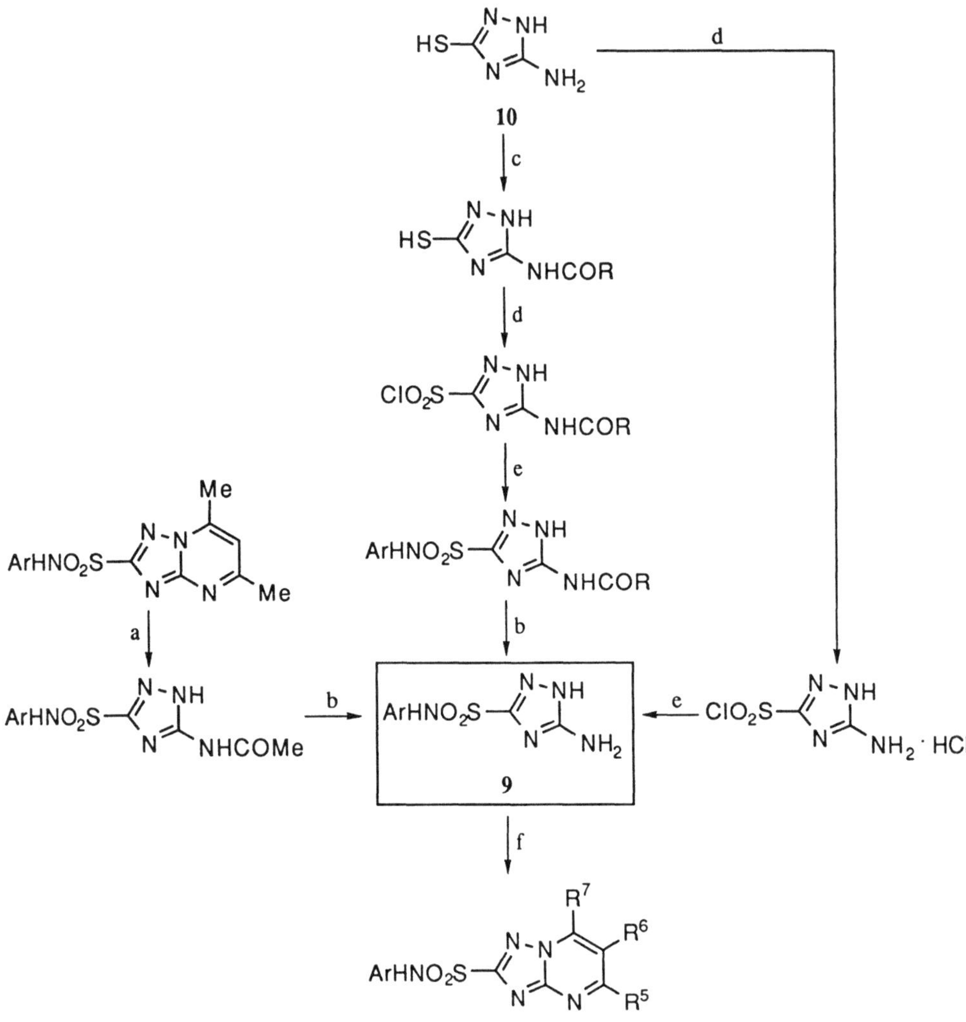

Scheme 5. (a) H_2O_2, KOH, H_2O; (b) HCl, H_2O or NaOH, H_2O; (c) $(RCO)_2O$ or RCOCl, pyridine; (d) Cl_2, H_3O^+; (e) $ArNH_2$, base; (f) $R^5COCHR^6COR^7$

chloride is added to consume the methyl mercaptide that is liberated. The preference for these ring systems to undergo nucleophilic substitution at the 7-position over the 5-position is a common element used to construct diverse substitution patterns in 1,2,4-triazolo[1,5-*a*]pyrimidines.

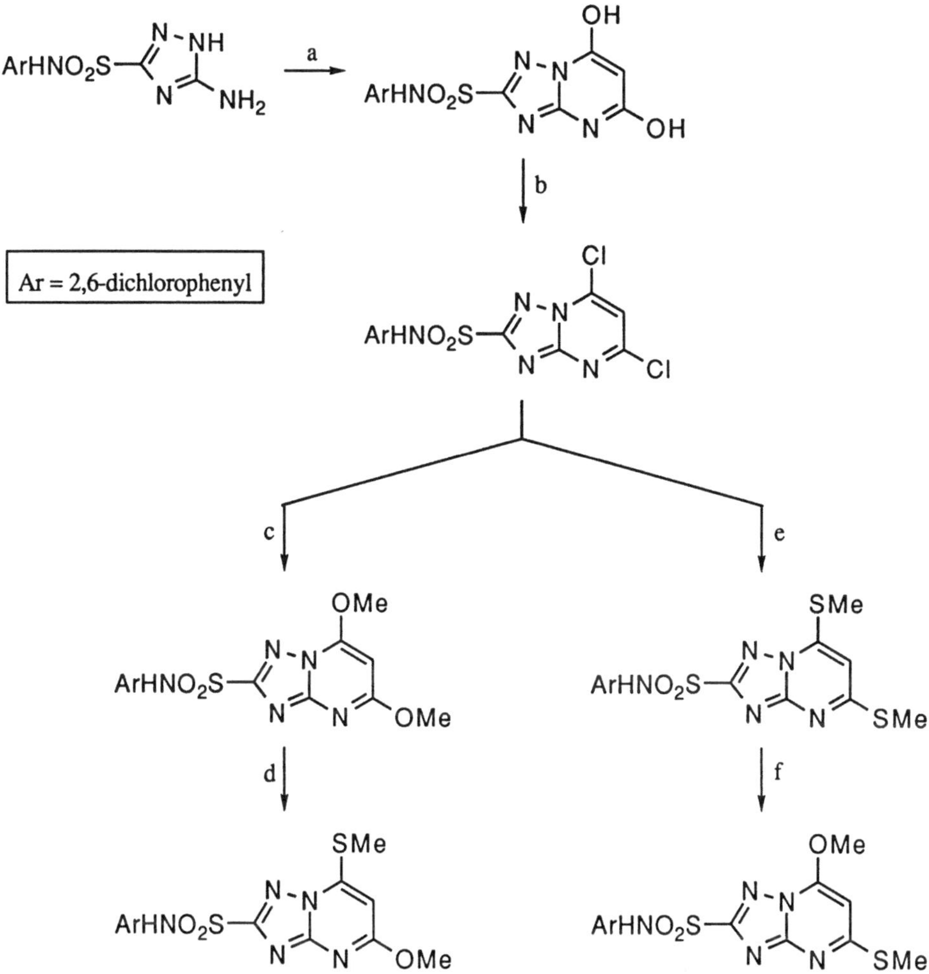

Scheme 6. (a) $CH_2(COOMe)_2$; (b) $POCl_3$; (c) NaOMe, MeOH, 0 °C-room temperature; (d) NaSMe, MeOH, 0 °C-room temperature; (e) NaSMe, MeOH; (f) NaOMe, BnCl, MeOH

2.3 1,2,4-Triazolo[1,5-*c*]pyrimidine-2-sulfonanilides

The first reported synthetic route to the 1,2,4-triazolo[1,5-*c*]pyrimidine sulfonanilides (**11**) is illustrated in Scheme 7 [40]. The exact details of some of the transformations and the origin of some of the intermediates in the route are vague, but the route appears to be applicable to the synthesis of alkyl substituted analogs (i.e. R^5, R^7 and/or R^8 = alkyl) and analogs containing hydroxy,

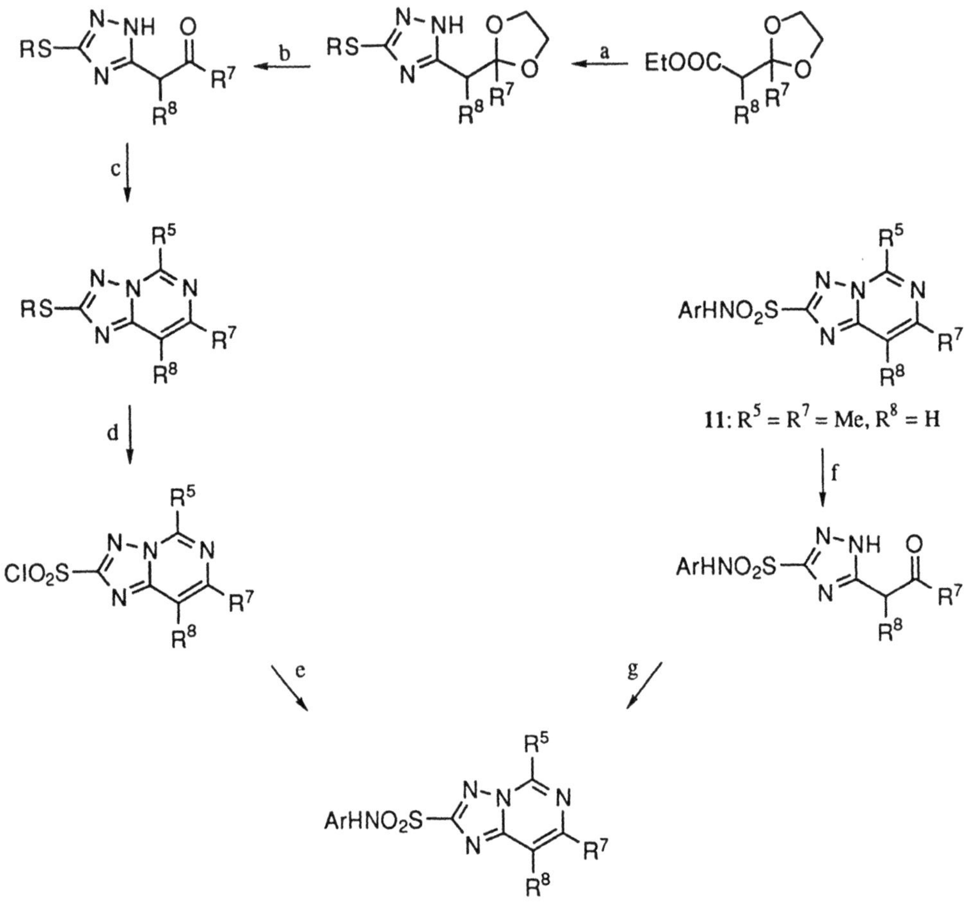

Scheme 7. (a) $H_2NNHC(=S)NH_2$, NaOEt, EtOH; (b) R = Bn: (1) NaOEt, EtOH, BnCl, (2) HCl (aq), reflux or R = H: HCl (aq), reflux; (c) R^5CONH_2; (d) Cl_2, H_3O^+; (e) $ArNH_2$, pyridine, CH_2Cl_2; (f) NaOH (aq), reflux; (g) R^5CONH_2 or $H_2NC(=X)NH_2$ (X = O, S or NH)

mercapto and amino substituents in the 5-position (i.e. R^5 = OH, SH or NH_2). The key intermediates for the synthesis of the 1,2,4-triazolo[1,5-*c*]pyrimidine ring are formed by direct synthesis of the triazole ring or by base catalyzed degradation of a 5,7-dimethyl substituted 1,2,4-triazolo[1,5-*c*]pyrimidine ring. Also, intermediates from the route are used to prepare compounds of structure **12** which have herbicidal activity (Scheme 8).

Another route to **11**, which affords a greater degree of flexibility, is illustrated in Scheme 9 [41, 42]. Nucleophilic substitution of a halogen at the 4- or 6-positions of a pyrimidine ring with hydrazine gives an intermediate which is

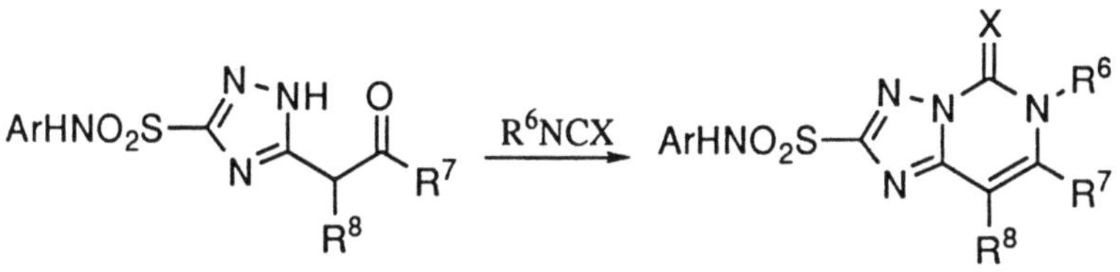

12

Scheme 8.

cyclized with carbon disulfide in the presence of base, followed by reaction with benzyl chloride to form a 1,2,4-triazolo[4,3-*c*]pyrimidine ring (i.e. **13**) containing a benzylthio group at the 3-position. The triazolo[4,3-*c*]pyrimidine undergoes a Dimroth rearrangement to the corresponding triazolo[1,5-*c*]pyrimidine (i.e. **14**) with base (e.g. NaOMe). Often it is unnecessary to isolate **13** as it can be converted to **14** in good yield in situ if a slight excess of base is used. The conversion of the benzylsulfide **14** to the corresponding sulfonyl chloride is accomplished with chlorine in aqueous acid. The reaction of the sulfonyl chloride with the appropriate aniline or *N*-trimethylsilyl aniline affords the sulfonanilide **11**. This route is applicable to the synthesis of sulfonanilides **11** which contain a variety of alkyl, alkoxy and halogen substituents.

While the route depicted in Scheme 9 is straightforward and very general, certain variations on the basic theme of the route are worthy of note. If the triazolo[4,3-*c*]pyrimidine contains an alkylthio group in the 5-position (i.e. **13**: R^5 = alkylthio), and it is treated with an alkali metal alkoxide in the presence of

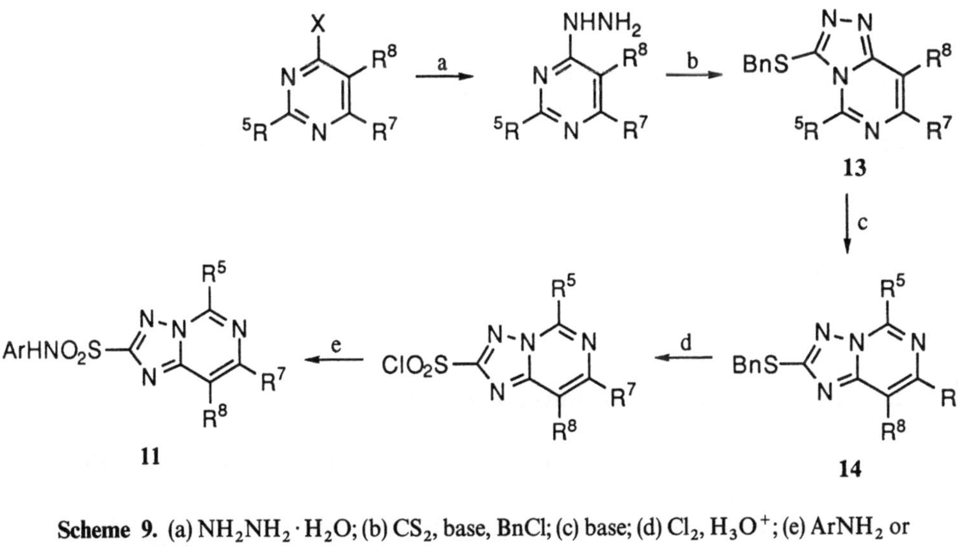

Scheme 9. (a) $NH_2NH_2 \cdot H_2O$; (b) CS_2, base, BnCl; (c) base; (d) Cl_2, H_3O^+; (e) $ArNH_2$ or ArNHTMS, base

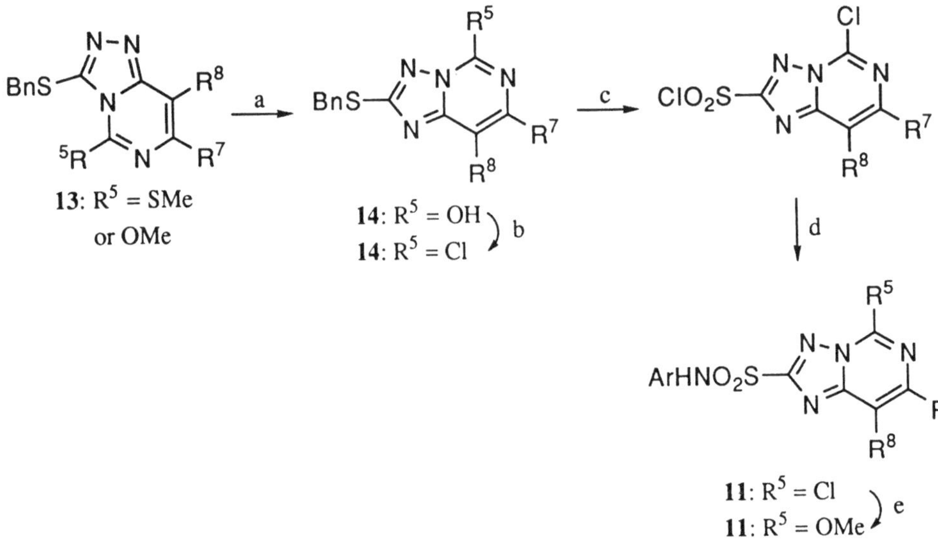

Scheme 10. (a) base; (b) $POCl_3$, MeCN, reflux; (c) Cl_2, H_3O^+; (d) $ArNH_2$ or ArNHTMS, base; (e) NaOMe, MeOH

a Michael acceptor (e.g. diethyl maleate), it yields the 5-alkoxy triazolo[1,5-*c*]pyrimidine (i.e. **14**: R^5 alkoxy) by displacement of the alkylthio group during the rearrangement. The Michael acceptor is necessary to consume the alkyl mercaptide that is liberated. In the absence of a Michael acceptor the reaction proceeds, but the alkoxy group is cleaved to yield a 5-hydroxy substituted triazolo[1,5-*c*]pyrimidine. This is a general method for introducing an alkoxy group at the 5-position of **11**.

The products from cleavage of ether substituents in the 5-position of a 1,2,4-triazolo[1,5-*c*]pyrimidine ring (i.e. **14**: R^5 = OH) are utilized to access 5-halo and 5-alkoxy substituted triazolo[1,5-*c*]pyrimidine sulfonanilides **11** (Scheme 10). Compound **14** (R^5 = OH) is converted to the corresponding chloro compound (**14**: R^5 = Cl). The chloro compound is transformed into the triazolo[1,5-*c*]pyrimidine sulfonanilide (**11**: R^5 = Cl) under standard conditions. The sulfonanilide **11** (R^5 = Cl) is reacted with nucleophiles (e.g. NaOMe) to form products resulting from substitution at the 5-position (e.g. **11**: R^5 = OMe).

2.4 Imidazolo[1,2-*a*]pyrimidine-2-sulfonanilides

The synthetic route to the imidazolo[1,2-*a*]pyrimidine-2-sulfonanilides (**15**) is illustrated by the example in Scheme 11 [43, 44]. Condensation of sodium 2-aminoimidazole-4-sulfonate with β-dicarbonyl compounds proceeds as described above for the triazolo[1,5-*a*]pyrimidine synthesis in terms of scope of

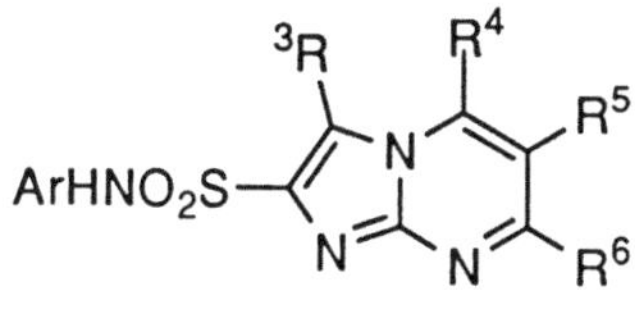

15

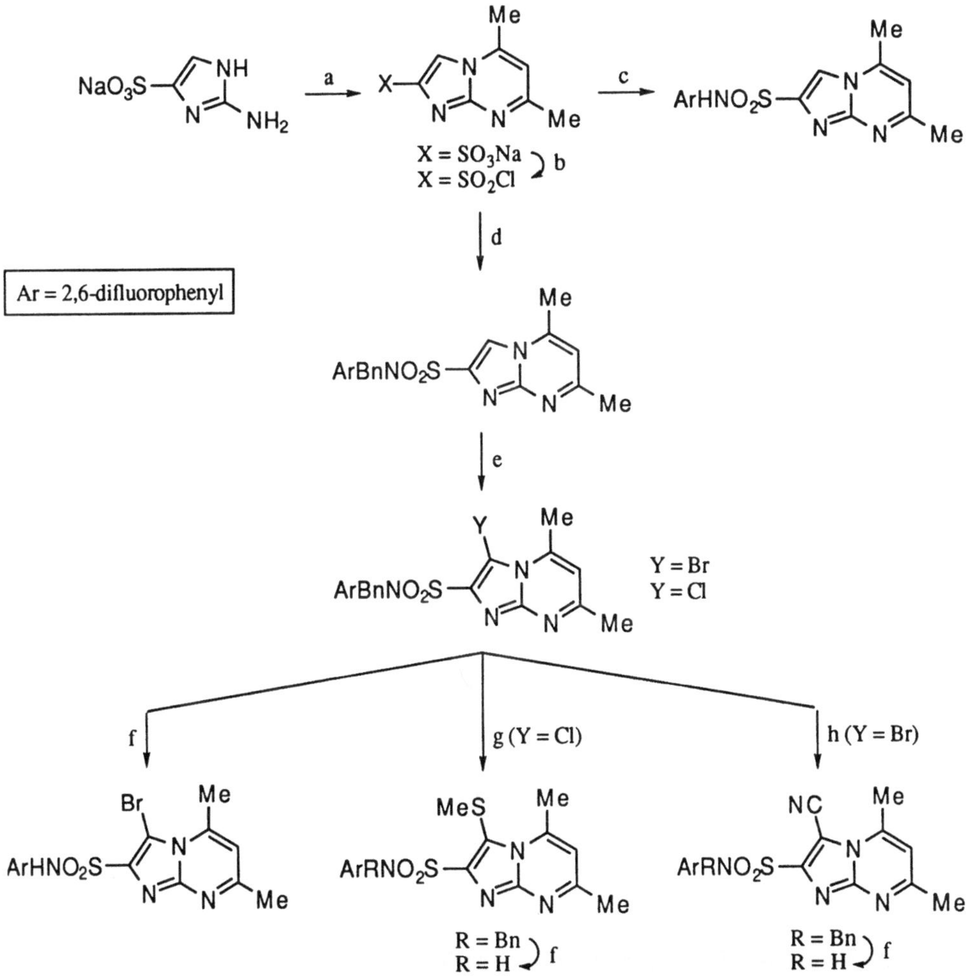

Scheme 11. (a) MeCOCH$_2$COMe, AcOH; (b) POCl$_3$, PhNEt$_2$ (cat.); (c) 2,6-difluoroaniline, pyridine; (d) *N*-benzyl-2,6-difluoroaniline, pyridine; (e) NCS, MeCN (Y = Cl) or NBS, MeCN (Y = Br); (f) AlCl$_3$, CH$_2$Cl$_2$; (g) NaSMe, DMF; (h) CuCN, *N*-methylpyrrolidone

reactivity and regiochemical outcome. The imidazolo[1,2-*a*]pyrimidine-2-sulfonic acid is converted to the corresponding sulfonyl chloride with phosphorus oxychloride in the presence of a catalytic quantity of *N*,*N*-diethylaniline. Reaction with anilines or *N*-alkylanilines (i.e. *N*-benzylanilines) is straightforward. The products from reaction with *N*-benzylanilines are further elaborated with the benzyl group serving as a protecting group which can be removed at the end of the synthesis. For example, the 3-position of the imidazolo[1,2-*a*]-pyrimidine ring is halogenated by reaction with *N*-halosuccinimides (e.g. NCS or NBS). The 3-halo derivatives undergo reactions with nucleophiles (e.g. NaSMe or CuCN) to displace the halogen. The benzyl group is easily removed with aluminum trichloride.

2.5 Pyrazolo[1,5-*a*]pyrimidine-2-sulfonanilides

The synthetic route to the pyrazolo[1,5-*a*]pyrimidine-2-sulfonanilides (**16**) is illustrated by the representative example depicted in Scheme 12 [12, 45–47]. The appropriate 3-amino-5-benzylthiopyrazole is prepared in two steps from ethyl 3,3-(bis)benzylthio-2-cyanoacrylate. Substituents (e.g. chlorine) are introduced at the 4-position of the pyrazole ring by electrophilic substitution. By analogy to condensations with the other aminoazoles described above, condensation with β-dicarbonyl compounds (e.g. β-ketoesters) proceeds smoothly to assemble the pyrazolo[1,5-*a*]pyrimidine ring. Also by analogy to the other azolopyrimidine ring systems, substituents on the ring (e.g. hydroxyl groups at the 7-position) can be manipulated to introduce functionality before or after formation of the sulfonanilide. All of the principles regarding control of substitution patterns via regioselective annulation, using unsymmetrical β-dicarbonyl compounds and selective manipulation of substituents on the ring as demonstrated for the triazolo[1,5-*a*]pyrimidine ring system, are applicable to these pyrazolo[1,5-*a*]pyrimidine ring systems.

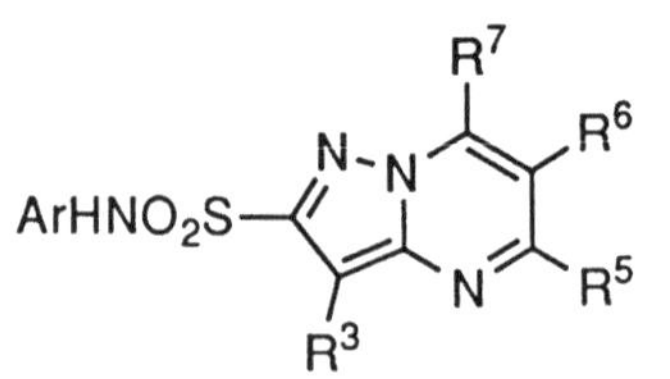

16

2.6 Triazolo[1,5-*a*]-1,3,5-triazine-2-sulfonanilides

There are a few reports of herbicidal activity associated with sulfonanilides based on the triazolo[1,5-*a*]-1,3,5-triazine ring system (**17**) [49, 50]. The direct synthesis of **17** is difficult due to the extremely electron deficient nature of the bicyclic ring

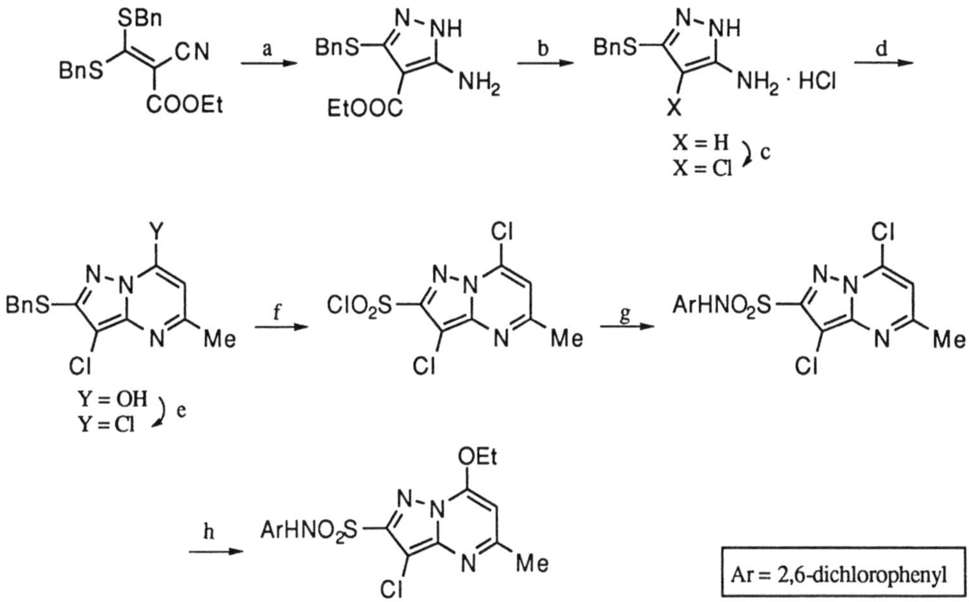

Scheme 12. (a) $NH_2NH_2 \cdot H_2O$, THF-EtOH; (b) (1) NaOH, (2) HCl; (c) Cl_2, CH_2Cl_2; (d) $MeCOCH_2COOEt$, AcOH; (e) $POCl_3$; (f) Cl_2, HCl (aq)-CH_2Cl_2; (g) 2,6-difluoro-*N*-trimethylsilylaniline, DMSO (cat.), MeCN; (h) NaOEt, EtOH

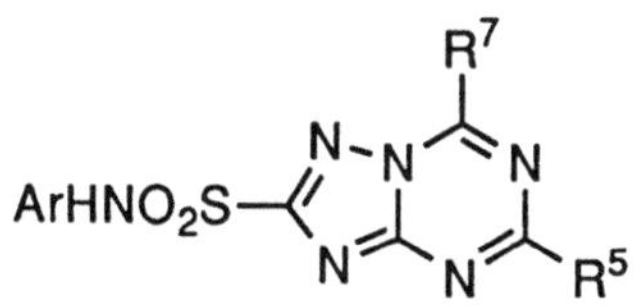

17

system and its tendency to undergo decomposition via ring opening reactions. The synthesis of this class of compounds is best illustrated by the two routes shown for the example of the syntheses of a more stable triazolo[1,5-*a*]-1,3,5-triazin-7-one-2-sulfonanilide (**18a**) in Scheme 13. The routes differ in regard to the stage in the synthesis at which the bicyclic ring system is formed. Triazolo[1,5-*a*]-1,3,5-triazine-7-thione-2-sulfonanilide (**18b**) can be synthesized by this route when an isothiocyanate is substituted in the synthetic scheme.

2.7 Pyrazolo[1,5-*a*]-1,3,5-triazine-2-sulfonanilides

The synthesis of the pyrazolo[1,5-*a*]-1,3,5-triazine-2-sulfonanilides (**19**) shares some of the features of the synthesis of the pyrazolo[1,5-*a*]pyrimidine-2-sulfonanilides (**8**) and the 1,2,4-triazolo[1,5-*a*]-1,3,5-triazine-2-sulfonanilides (**17**)

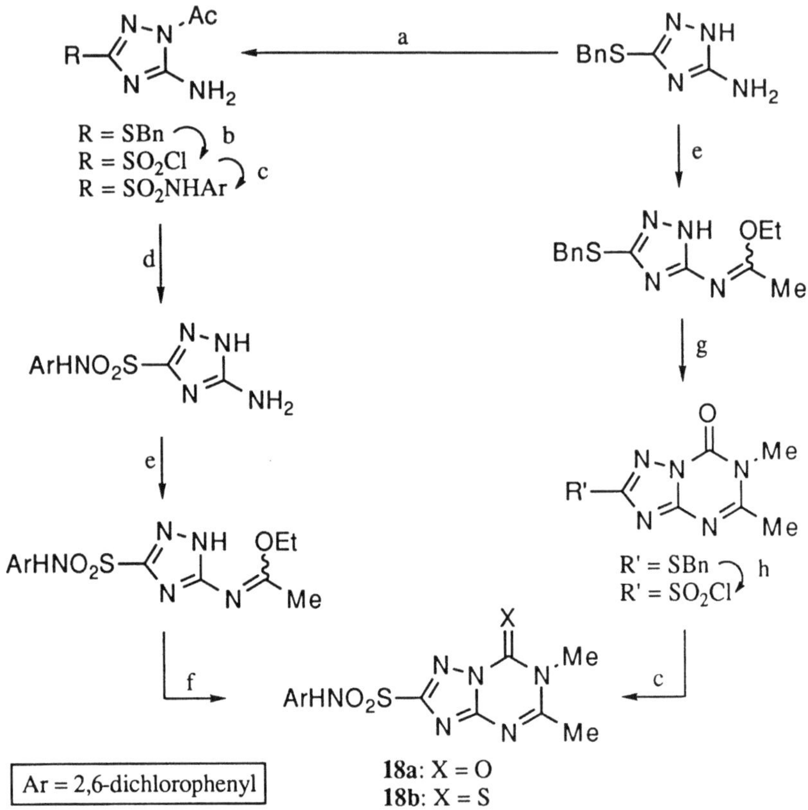

Scheme 13. (a) AcCl, Et_3N, CH_2Cl_2; (b) Cl_2, AcOH-H_2O -10 to 0 °C; (c) 2,6-dichloroaniline, pyridine; (d) NaOH (aq.); (e) $(EtO)_3CMe$, AcOH, MeCN, reflux; (f) MeNCX, Et_3N, THF; (g) MeNCO, Et_3N, THF; (h) Cl_2, AcOH-H_2O, 5 to 10 °C

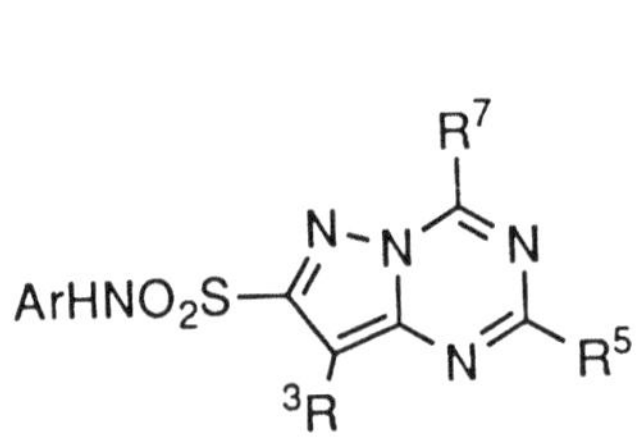

19

[51]. Representative examples of the synthetic routes are illustrated in Scheme 14. One route involves the formation of the bicyclic ring system prior to formation of the sulfonamide bond. It is only applicable to the synthesis of **20a** due to the loss of sulfur at the 7-position during the conversion of the benzyl sulfide to the sulfonyl chloride. The second route, in which the sulfonamide bond is formed prior to construction of the bicyclic ring system, can be applied to the synthesis of both **20a** and **20b**.

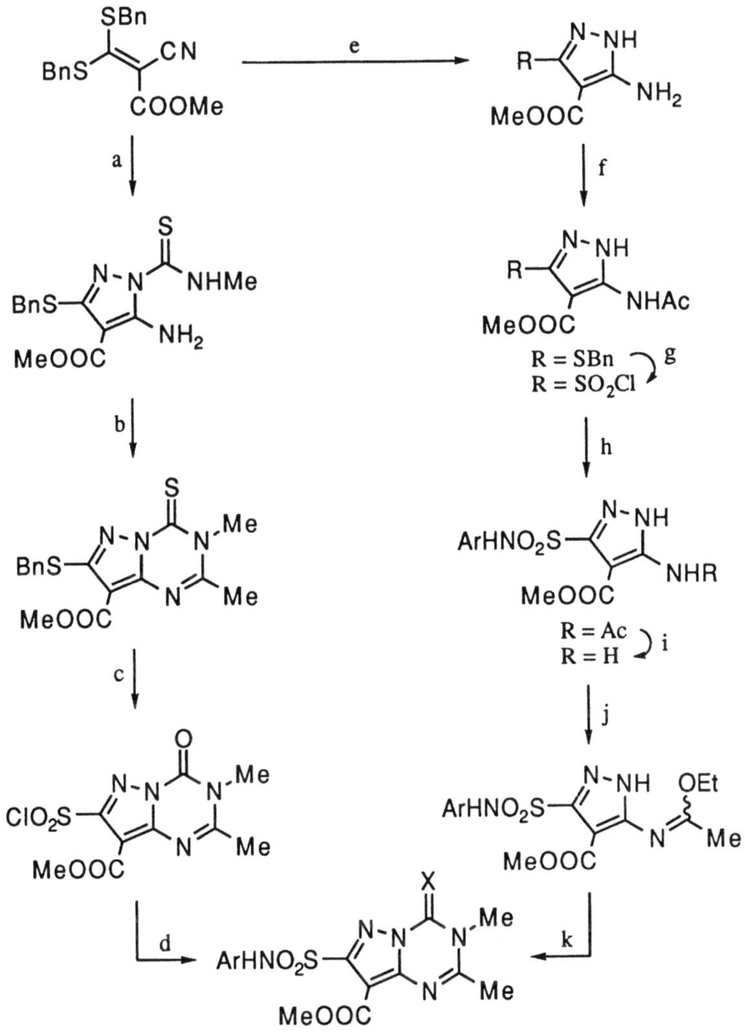

20a: X = O
20b: X = S

Scheme 14. (a) $H_2NHC(=SH)NH_2$, EtOH, reflux; (b) $MeC(OEt)_3$, reflux; (c) Cl_2, AcOH-H_2O, 10 °C; (d) 2,6-dichloroaniline, pyridine; (e) 1) *t*-$BuOCONHNH_2$, EtOH, reflux; 2) AcOH, 110 °C; (f) Ac_2O, AcOH, reflux; (g) SO_2Cl_2, CH_2Cl_2, 0 °C; (h) 2,6-difluoroaniline, pyridine; (i) HCl, MeOH, reflux; (j) $MeC(OEt)_3$, AcOH, MeCN, reflux; (k) MeCNX, Et_3N, THF

2.8 Thiazolo[3,2-*b*]-1,2,4-triazole-2-sulfonanilides

The thiazolo[3,2-*b*]-1,2,4-triazole-2-sulfonanilides (**21**) are the only examples of a 5,5-fused ring systems in this class of herbicides. Chemists at Schering reported the first synthesis of **21** [52]. The synthetic route is illustrated with the example in

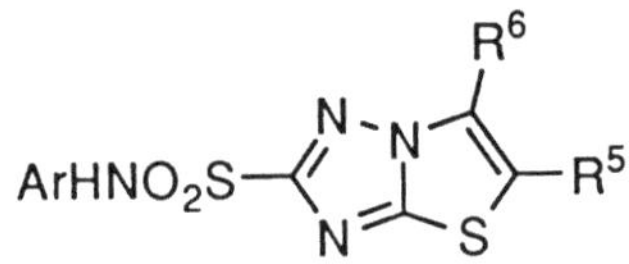

21

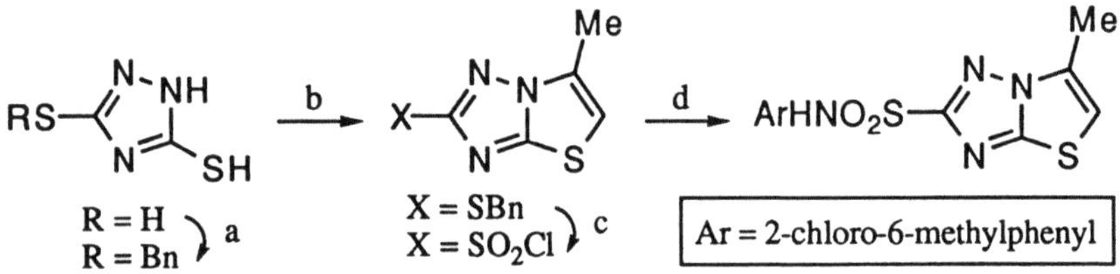

Scheme 15. (a) NaOMe, MeOH, BnCl; (b) $ClCH_2COMe$, EtOH, reflux; (c) Cl_2, AcOH-H_2O, < 5 °C or SO_2Cl_2, CH_2Cl_2, 5–10 °C; (d) 2-chloro-6-methylaniline, pyridine

Scheme 15. The bicyclic ring system is constructed through sequential alkylation of one of the mercapto groups of 3,5-dimercapto-1,2,4-triazole with benzyl chloride, followed by alkylation of the other with an alpha halocarbonyl compound or a synthetic equivalent (e.g. an alpha haloacetal). Ring closure is accomplished in situ under conditions for the alkylation or after isolation of the alkylation product. Examples are provided for a number of analogs containing H, alkyl, aryl, alkoxycarbonyl and acyl groups in either the 5- or 6-positions (i.e. R^5 or R^6 in **21**).

3 Structure–Activity Relationships

3.1 (*N*-1,2,4-Triazolo[1,5-*a*]pyrimidine)benzene Sulfonamides

Some of the structure–activity trends for compounds of general structure **3** are described in detail in the literature [1]. For substitution on the phenyl ring, substitution at the ortho-position produces greater activity on broadleaf weed species than substitution at the meta- or para-positions. Substitution at the ortho-position with electron-withdrawing groups (i.e. NO_2, CF_3 or Cl) produces greater activity than substitution with electron donating groups. Comparing the activity of analogs on the enzyme leads to the same general conclusions, although the exact ranking of analogs by activity may be slightly different. For disubstituted phenyl rings, the 2,6-disubstitution exhibits both herbicidal activity and ALS inhibition that are far superior to all other isomeric substitution patterns.

General trends in structure–activity relationships as a function of changes in triazolopyrimidine ring substitution of **3** are less straightforward. Higher levels of herbicidal activity and ALS inhibition are associated with alkyl (e.g. methyl) or alkoxy (e.g. methoxy and ethoxy) substitution in the 5- and 7-positions (R^5 and R^7). Lower levels of activity are associated with halo or haloalkyl substitution.

3.2 1,2,4-Triazolo[1,5-*a*]pyrimidine-2-sulfonanilides

Comprehensive accounts of many of the structure–activity trends in the 1,2,4-triazolo[1,5-*a*]pyrimidine-2-sulfonanilides (**8**) exist in the literature [1, 28, 31, 33]. The trends in in vivo activity against broadleaf weeds and in vitro activity against ALS are similar for most compounds **8** where Ar is equal to substituted phenyl. In general *ortho* substituted analogs are considerably more active than their *meta* or *para* counterparts. Compounds with electron-withdrawing groups in the *ortho* position are more active than those with electron-donating groups in the same position. Substitution at the 2- and 6-positions on the phenyl ring produces much higher levels of activity than all other di-substituted isomers. The highest levels of activity for 2,6-disubstituted analogs are associated with compounds in which one of the substituents is electron-withdrawing and the other substituent is electron-withdrawing or methyl or methoxy. Some 2,3,6,5-trisubstituted phenyl analogs demonstrate very high levels of activity (e.g. **8**: Ar = 2,6-dichloro-3-methylphenyl or 2,3-dimethyl-6-nitrophenyl). Much of the activity as a function of phenyl substitution of **8** is determined by the contribution which the substitution makes to modulating the pK_a of the sulfonamide and the physical properties (i.e. log P) of the molecule. As such, changes in Ar within this series of compounds (**8**) often have more pronounced effects than similar changes in **3**, presumably due to the greater impact of the changes on the pK_a of **8**.

As in the case of **3** trends in activity of **8** as a function of substitution on the triazolopyrimidine ring are not well understood [1, 28, 31, 33]. In general, high levels of in vitro and in vivo activity are found with compounds possessing alkyl (e.g. methyl) and/or alkoxy (e.g. methoxy or ethoxy) substitution at the 5- and/or 7-positions (R^5 and R^7). Lower activity is associated with alkylthio, amino, aryl, carboxyl and halo substitution.

The soil behavior of compounds of general structure **8** is influenced by changes in the substitution patterns on the molecule [28, 33]. Changes in substitution on Ar have only a modest impact on the soil half-life of the compounds with the exception of compounds containing ester functionality. These materials possess shorter soil half-lives than other closely related analogs presumably because of hydrolysis of the ester to the inactive carboxylic acid. By contrast substitution on the triazolo[1,5-*a*]pyrimidine ring impacts the soil half-life of **8** to a great extent. Alkoxy substitution generally lowers the half-life of these materials in soil [33].

3.3 1,2,4-Triazolo[1,5-*c*]pyrimidine-2-sulfonanilides

The structure–activity relationships for the 1,2,4-triazolo[1,5-*c*]pyrimidine-2-sulfonanilides (**11**) have been studied extensively [40–42]. The trends in in vitro and in vivo activity as a function of substitution on Ar (where Ar = substituted phenyl) are analogous to those described previously for the triazolo[1,5-*a*]pyrimidine-2-sulfonanilides (**8**).

Table 1 presents a compilation of herbicidal activity and ALS inhibition data for a representative series of analogs (**11**) with varying substitution on the heterocyclic ring system. The broadleaf weed activity is expressed as an average of the concentration required for 80% growth reduction (GR_{80}) for five to eight species chosen from among the following: *Abutilon theophrasti*, *Amaranthus retroflexus*, *Chenopodium album*, *Datura stramonium*, *Ipomoea* spp., *Polygonum convolvulus*, *Veronica persica* and *Xanthium strumarium*. The grass weed activity is expressed as an average of the concentration required for 50% growth reduction (GR_{50}) for five species chosen from among the following: *Alopecurus myosuroides*, *Avena fatua*, *Digitaria sanguinalis*, *Echinochloa crus-galli*, *Setaria* spp. and *Sorghum halepense*. The enzyme inhibition data is determined using ALS isolated from etiolated barley tissue. The data indicate that alkoxy

Table 1. Biological activity for **11** (Ar = 2,6-dichlorophenyl)

R^5	R^7	R^8	Avg. GR_{80} (ppm) Broadleaf Weeds Postemergence	Avg. GR_{50} (ppm) Grass Weeds Postemergence	I_{50} (μM) ALS
Me	Me	H	200	1000	0.31
Me	Cl	H	400	940	0.52
Me	CF_3	H	480	1200	0.43
Me	OMe	H	>500	>1000	2.2
OMe	H	H	1.2	2.5	0.021
OMe	Me	H	4.8	5.4	0.0012
OMe	Et	H	100	51	0.012
OMe	F	H	23	2.3	0.0032
OMe	Cl	H	4.3	10	0.0022
OMe	CF_3	H	46	45	0.0090
OMe	CH_2OMe	H	37	19	0.26
OMe	NMe_2	H	170	>250	0.027
OEt	F	H	4.8	11	0.0016
O-*n*-Pr	F	H	26	⩾16	0.017
O-*i*-Pr	F	H	10	⩾16	0.065
OMe	H	F	2.0	4.3	0.0038
OMe	H	Cl	1.4	2.8	0.0012
OMe	H	CF_3	200	>250	0.14
OMe	H	Me	16	22	0.11
OMe	H	OMe	9.7	13	0.0064
OMe	H	SMe	70	160	0.014

Table 2. Soil decomposition of **11** (Ar = 2,6-dichlorophenyl)

R^5	R^7	R^8	Approximate Soil Half-Life (days)[a]
Me	Me	H	180
OMe	OMe	H	8
OMe	Me	H	21
OMe	Cl	H	12
OMe	F	H	3
Me	OMe	Cl	9

[a] Greenhouse bioassay using sunflower (sandy loam soil, 1% OM, pH 7.2)

substitution at the 5-position (R^5) is superior to alkyl substitution, with methoxy and ethoxy substituted analogs having comparable inherent activity and *n*-propoxy and *i*-propoxy being less active. Substitution at the 7- and 8-positions (R^7 and R^8) with both electron-withdrawing and electron-donating substituents produces highly active compounds. Combinations of alkoxy substitution at the 5-position and halogen, alkyl or alkoxy substitution at the 7- or 8-position give rise to the highest levels of activity.

Soil decomposition data for a series of triazolo[1,5-*c*]pyrimidine-2-sulfonanilide analogs (**11**) with varying heterocyclic substitution are shown in Table 2. All of the compounds containing methoxy substitution at the 5- or 7-positions (R^5 and R^7) have significantly shorter soil half-lives than the 5,7-dimethyl substituted analog (**11**; $R^5 = R^7 =$ Me). This trend in the impact of alkoxy and alkyl substitution on the bicyclic heterocycle is analogous to that observed in the triazolo[1,5-*a*]pyrimidine sulfonanilides (**8**).

3.4 Imidazolo[1,2-*a*]pyrimidine-2-sulfonanilides

The structure–activity relationships for the imidazolo[1,2-*a*]pyrimidine-2-sulfonanilides (**15**) are analogous to those for the 1,2,4-triazolo[1,5-*a*]-pyrimidine-2-sulfonanilides (**8**) with respect to substituents on the aryl ring (Ar) and substituents on the pyrimidine portion of the bicyclic heterocycle (i.e. R^4, R^5 and R^6 in **15** are equivalent to R^7, R^6 and R^5 in **8** respectively) [43,44]. Table 3 illustrates the trends in activity of **15** on broadleaf (*Abutilon theophrasti, Amaranthus retroflexus, Datura stramonium, Ipomoea* spp. and *Xanthium strumarium*) and grass (*Avena fatua, Digitaria sanguinalis, Echinochloa crus-galli, Setaria* spp. and *Sorghum halepense*) weed species and on inhibition of ALS from barley as a function of substitution at the 3-position (R^3). Compounds substituted with electron-withdrawing groups in the 3-position are more active in vitro and in vivo than compounds substituted with electron-donating groups. All

Table 3. Biological activity of **15** (Ar = 2,6-difluorophenyl, $R^4 = R^6 = Me$, $R^5 = H$)

R^3	Average GR_{80} on Broadleaf Weeds Postemergence (ppm)	Average GR_{80} on Grass Weeds Postemergence (ppm)	ALS I_{50} (μM)
H	1400	≫2000	24
Cl	160	910	1.8
Br	420	1200	–[a]
SMe	≫2000	≫2000	15
CN	250	≫2000	2.1

[a] Value not determined. The I_{50} values (μM) for **15** (Ar = 2,6-difluorophenyl, $R^3 = Br$, $R^4 = R^6 = OMe$, $R^5 = H$) and **15** (Ar = 2,6-difluorophenyl, $R^3 = H$, $R^4 = R^6 = OMe$, $R^5 = H$) are 0.13 and 1.1 respectively

compounds are more active on broadleaf weeds than grass weeds. The compounds in Table 3 are also less active than a representative 1,2,4-triazolo[1,5-*a*]pyrimidine-2-sulfonanilide (**8**: Ar = 2,6-difluorophenyl, $R^5 = R^7 = Me$, $R^6 = H$), which exhibited GR_{80} values of 40 and 200 ppm on broadleaf and grass weeds respectively and an I_{50} of 0.51 μM against ALS. Approximate soil half-lives of 59 and 64 days respectively were determined for the 3-chloro and 3-bromo analogs (Table 3). These values are considerably lower than the value of 171 days for a representative 1,2,4-triazolo[1,5-*a*]pyrimidine-2-sulfonanilide (**8**: Ar = 2,6-dichlorophenyl, $R^5 = R^7 = Me$, $R^6 = H$).

3.5 Pyrazolo[1,5-*a*]pyrimidine-2-sulfonanilides

As was discussed above for the imidazolo[1,2-*a*]pyrimidine-2-sulfonanilides (**15**), the structure–activity relationships for the pyrazolo[1,5-*a*]pyrimidine-2-sulfonanilides (**16**) are analogous to those for the 1,2,4-triazolo[1,5-*a*]-pyrimidine-2-sulfonanilides (**8**) with respect to substituents on both the aryl ring (Ar) and the pyrimidine portion of the bicyclic heterocycle (i.e. R^7, R^6 and R^5 in **16** are equivalent to R^7, R^6 and R^5 in **8** respectively) [12, 45–48]. A comparison of postemergence test data for a series of compounds (**16**: Ar = 2,6-difluorophenyl, $R^5 = R^7 = Me$, $R^6 = H$) in which R^3 is varied reveals an order of broadleaf weed (*Abutilon theophrasti*, *Amaranthus retroflexus*, *Datura stramonium*, *Ipomoea* spp. and *Xanthium strumarium*) activity of Cl > CN > COOEt > NH_2 ≫ H, CF_3 (Table 4). As was observed for substitution at the 3-position of **15**, electron-withdrawing substituents produce the highest levels of activity. One notable exception is CF_3 which was almost completely devoid of postemergence activity at concentrations up to 1000 ppm. Similar trends were observed in vitro against ALS from barley (Table 4). The most active analog in this comparison ($R^3 = Cl$) exhibited GR_{50} values of 5.0 and 45 ppm on broadleaf and grass weeds respectively. The approximate soil half-life for a representative analog (**16**: Ar = 2,6-difluorophenyl, $R^3 = Cl$, $R^5 = R^7 = Me$, $R^6 = H$) was

Table 4. Biological activity of **16** (Ar = 2,6-dichlorophenyl, $R^4 = R^6 = Me$, $R^5 = H$)

R^3	Average GR_{50} on Broadleaf Weeds Postemergence (ppm)	ALS I_{50} (μM)
H	inactive at 4000	30
CN	220	2.3
COOEt	710	10
Cl	5.0	0.86
CF_3	>1000	11
COOH	not determined	240
NH_2	>500	4.8
NMe_2	$\gg 250$	2.0

determined to be 55 days. This value is shorter than values for analogous 1,2,4-triazolo[1,5-*a*]pyrimidine-2-sulfonanilides (**8**) and comparable to values for analogous imidazolo[1,2-*a*]pyrimidine-2-sulfonanilides (**15**) (see above). It is interesting to note that one analog (**16**: Ar = 2,6-dichlorophenyl, $R^3 = CONH_2$, $R^5 = Me$, $R^6 = R^7 = H$) with the same substitution on the pyrimidine fragment as flumetsulam (**1**) also exhibits weed control without injury to soybean in greenhouse tests [48].

3.6 Other Structures

Very limited data is available regarding the structure–activity trends in the triazolo[1,5-*a*]-1,3,5-triazine-2-sulfonanilides (**17**) or the pyrazolo[1,5-*a*]-1,3,5-triazine-2-sulfonanilides (**19**) [49–51]. As has been the case with other ring systems, changes in biological activity with changes in Ar or substitution on Ar appear to follow well established patterns. On the bicyclic heterocycle, the best activity is observed with alkyl substitution (i.e. R^5 and R^6 = alkyl, especially methyl) in **18a**, **18b**, **20a** and **20b**. One analog (**18b**: Ar = 2,6-dichlorophenyl) gave complete control of *Setaria, Helianthus, Stellaria, Abutilon, Viola, Brassica, Solanum, Matricaria* and *Echinochloa* in preemergence applications at 0.1 kg/ha [49]. Substitution at the 3-position (R^3) in **19** seems to follow a pattern of herbicidal activity in which COOMe ≈ COOEt > $CONH_2$ > SO_2Me [51].

Only a limited amount of data describing the herbicidal activity of the thiazolo[3,2-*b*]-1,2,4-triazole-2-sulfonanilides (**21**) is reported in the literature [48, 52]. This data indicates that trends in structure activity relationships as a function of Ar and substitution on Ar are similar to those described above for other compounds (e.g. **8**). The data suggests that the order of herbicidal activity as a function of substitution on the bicyclic heterocycle (R^5, R^6) is as follows: H, Me ≥ Me, H > Me, Me ≈ H, Et ≈ H, CH_2Cl. Substitution at R^5 or R^6 with a halogen appears to produce good activity [48]. One analog (**21**: Ar = 2-chloro-6-fluorophenyl, $R^5 = R^6 = H$) was shown to possess a longer soil half-life and greater soil mobility than atrazine [48].

4 Conclusions

The triazolopyrimidine class of ALS inhibiting herbicides has grown to include a variety of structures containing bicyclic heterocycles containing bridgehead nitrogen atoms. Members of the class have demonstrated a high degree of herbicidal efficacy in controlling weeds in major agronomic crops. This herbicidal efficacy is often accompanied by favorable toxicological profiles and desirable environmental properties. Representative compounds from the class (e.g. **1** and **2**) offer utility which often complements that of compounds from other structural classes of ALS inhibiting herbicides.

5 References

1. Kleschick WA, Costales MJ, Dunbar JE, Meikle RW, Monte WT, Pearson NR, Snider SW, Vinogradoff AP (1990) Pestic Sci 29: 341
2. Kleschick WA, Gerwick BC, Carson CM, Monte WT, Snider SW (1992). J Ag Food Chem 40: 1083
3. Snel M, Watson P, Gray N, Kleschick W, Carson C (1993). In: 45th Symposium on Crop Protection, Ghent, Belgium, May 4 1993
4. Gerwick BC, Subramanian MV, Loney-Gallant VI, Chandler DP (1990). Pestic Sci 29: 357
5. Kleschick WA (1989). US Pat 4 822 404 (Dow)
6. Kleschick WA, Vinogradoff AP, Dunbar JE (1987). US 4 638 075 (Dow)
7. Kleschick WA, Vinogradoff AP, Dunbar JE (1987). US 4 650 892 (Dow)
8. Kleschick WA, Vinogradoff AP, Dunbar JE (1988). US 4 772 720 (Dow)
9. Westermann J, Krueger M, Arndt F (1987). DE 3 539 386 (Schering)
10. Kleschick WA, Dunbar JE, Snider SW, Vinogradoff AP (1988). J Org Chem 53: 3120
11. Pearson NR, Kleschick WA (1987). US 4 685 958 (Dow)
12. Hunt RG (1987). EP Appl 244 097 (Schering)
13. Tseng C-P (1983). US 4 411 690 (Du Pont)
14. Okada Y, Aoki I, Okajima N, Kuragano T (1987). EP Appl 239 064 (Takeda)
15. Bellina RF (1991). US 5 069 709 (Du Pont)
16. Okajima N, Aoki I, Kuragano T, Okada Y (1991). Pestic Sci 32: 91
17. Kleschick WA, Ehr RJ, Gerwick BC, Monte WT, Pearson NR, Costales MJ, Meikle RW (1988). US 4 740 223 (Dow)
18. Kleschick WA, Gerwick BC, Ehr RJ, Monte WT, Pearson NR, Meikle RW, Costales MJ (1988). US 4 741 764 (Dow)
19. Kleschick WA, Gerwick BC, Ehr RJ, Monte WT, Pearson NR, Meikle RW, Costales MJ (1988). US 4 755 212 (Dow)
20. Kleschick WA, Ehr RJ, Costales MJ, Gerwick BC, Meikle RW, Monte WT, Pearson NR (1989). US 4 818 273 (Dow)
21. Kleschick WA, Ehr RJ, Costales MJ, Gerwick BC, Meikle RW, Monte WT, Pearson NR (1989). US 4 886 883 (Dow)

22. Pearson NR, Kleschick WA, Carson CM (1990). US 4904301 (Dow)
23. Kleschick WA, Ehr RJ, Costales MJ, Gerwick BC, Meikle RW, Monte WT, Pearson NR (1990). US 4954163 (Dow)
24. Pearson NR, Kleschick WA, Carson CM (1990). US 4959473 (Dow)
25. Pearson NR, Carson CM, Kleschick WA (1990). US 4979981 (Dow)
26. Kleschick WA, Costales MJ, Ehr RJ, Gerwick BC, Meikle RW, Monte WT, Pearson NR (1991). US 4983772 (Dow)
27. Pearson NR, Carson CM, Kleschick WA (1991). US 5015286 (Dow)
28. Kleschick WA, Gerwick BC (1989). In: Copping LG, Dodge AD, Dalzeil J (eds) Prospects for Amino Acid Biosynthesis Inhibitors in Crop Protection and Pharmaceutical Chemistry. Lavenham Press Limited, Lavenham, Suffolk, pp 139–146 (BCPC Monograph No. 42)
29. Monte WT, Kleschick WA, Meikle RW, Snider SW, Bordner J (1989). J Heterocyclic Chem 26: 1393
30. Kleschick WA, Bordner J (1989). J Heterocyclic Chem 26: 1489
31. Kleschick WA, Costales MJ, Gerwick BC, Holtwick JB, Meikle RW, Monte WT, Pearson NR, Snider SW, Subramanian MV, VanHeertum JC, Vinogradoff AP (1992). In: Baker DR, Fenyes J, Steffens JJ (eds) Synthesis and Chemistry of Agrochemicals III. American Chemical Society, Washington, DC, pp 10–16 (ACS Symposium Series 504)
32. Jelich K, Santel H-J, Schmidt RR, Strang H (1988). EP Appl 268951 (Bayer)
33. Kleschick WA, Carson CM, Costales MJ, Doney JJ, Gerwick BC, Holtwick JB, Meikle RW, Monte WT, Little JC, Pearson NR, Snider SW, Subramanian MV, VanHeertum JC, Vinogradoff AP (1992). In: Baker DR, Fenyes J, Steffens JJ (eds) Synthesis and Chemistry of Agrochemicals III. American Chemical Society, Washington, DC, pp 17–25 (ACS Symposium Series 504)
34. McKendry LH (1990). US 4910306 (Dow)
35. Jelich K, Kramer W, Santel H-J, Schmidt RR, Strang RH (1988) EP Appl 256396 (Bayer)
36. Monte WT (1988). US 4734123 (Dow)
37. Little JC, Costales MJ, Shankar RB, Pews RG, Mariam KG, Thompson SD (1990). EP Appl 375060 (Dow)
38. Little JC, Thibos PA, Mariam KG, Thompson SD (1990). EP Appl 375061 (Dow)
39. Kim KK, Krauss RC, Orvik JA (1991). US 4988812 (DowElanco)
40. Green DE (1987). EP Appl 244948 (Schering)
41. VanHeertum JC, Gerwick BC, Kleschick WA (1991). US 5010195 (Dow)
42. VanHeertum JC, Gerwick BC, Kleschick WA, Johnson TC (1992). US 5163995 (DowElanco)
43. Pearson NR, Kleschick WA, Bartley SL (1988). US 4731446 (Dow)
44. Pearson NR, Bartley SL, Kleschick WA (1989). US 4799952 (Dow)
45. Vinogradoff AP, Kleschick WA (1991). US 4992091 (Dow)
46. Vinogradoff AP, Kleschick WA (1991). US 4997940 (Dow)
47. Vinogradoff AP, Kleschick WA (1991). US 5021591 (Dow)
48. Percival A (1991). Pestic Sci 31: 569
49. Pearson NR, Kleschick WA (1986). US 4605433 (Dow)
50. Westermann J, Kruger M, Arndt F, Rees R, Kotter C (1987). EP Appl 272628 (Schering)
51. Kruger M, Westermann J, Arndt F, Rees R, Hunt RG (1988). EP Appl 297490 (Schering)
52. Gates PS (1987). EP Appl 244098 (Schering)

Mechanisms of Tolerance to Triazolopyrimidine Sulfonanilide Herbicides

B. C. Gerwick, G. J. DeBoer, P. R. Schmitzer

DowElanco, Discovery Research, P.O. Box 68955, 9410 Zionsville Road, Indianapolis, IN 46268-1053, USA

Two mechanisms by which plants are resistant to triazolopyrimidine sulfonanilide herbicides are alterations in ALS target site sensitivity and metabolic detoxification. Examples of the latter include O-dealkylation and triazolopyrimidine ring cleavage in soybeans, alkyl oxidation in wheat and soybeans, and glucose conjugation in wheat. Differential rates of metabolism through common and species-specific pathways result in differential sensitivity at the whole plant level. Modifications in ALS can also impart tolerance. Evaluation of sulfonylurea and imidazolinone resistant weeds for sensitivity to triazolopyrimidines suggests greater likelihood of cross-resistance from sulfonylureas. A rapid method to identify target-site based resistance is described.

List of Symbols and Abbreviations

AHAS	acetohydroxyacid synthase
ALS	acetolactate synthase
C	Celsius
CPCA	1,1-cyclopropanedicarboxylic acid
$g\,ha^{-1}$	grams per hectare
gfw	gram fresh weight
GR_{80}	application rate of compound resulting in 80% reduction of growth
h	hour
I_{50}	concentration of inhibitor resulting in 50% inhibition
IM	imidazolinone
KARI	ketol-acid reductoisomerase
NA	not available
ppm	parts per million
R	resistant
R/S	ratio of activity in resistant divided by activity in susceptible
S	susceptible
SU	sulfonylurea
$T_{1/2}$	half-life
TP	triazolopyrimidine sulfonanilide
μg	microgram
μM	micromolar
$<$	less than
$>$	greater than
%	percent

1 Introduction

Herbicide resistance results when a plant has the ability to limit the toxic effect produced by a chemical. Resistance in plants can be attributed to a number of mechanisms such as reduced uptake and translocation, compartmentalization of the herbicide away from the active-site, active-site insensitivity, and metabolism of the herbicide. The most important mechanism for selectivity between crops and weeds is differential metabolism. Metabolism has been shown to produce selectivity in herbicide chemical classes including triazines, acetanilides, phenoxies, sulfonylureas, imidazolinones and numerous others [1, 2]. Although metabolism does play a part in the resistance of weeds to a herbicide, active-site insensitivity has been studied to a greater extent as a mode of resistance.

Three commercial areas of herbicidal chemistry, namely the sulfonylureas (SU), imidazolinones (IM) and triazolopyrimidine sulfonanilides (TP), act through inhibition of the enzyme acetolactate synthase (ALS/AHAS, EC 4.1.3.18) [3, 4, 5] which catalyzes the first committed step in the biosynthesis of leucine, isoleucine and valine. All three families of inhibitors have been demonstrated by Schloss et al. to compete for binding to the enzyme [6]. This suggests they share a common binding domain within the enzyme. Recent data also suggests that the binding domain of these inhibitors contains IM exclusive regions in addition to structurally shared elements for the SU and TP inhibitors [7, 8].

Differential metabolism has been demonstrated in crops for all three classes of ALS-inhibiting herbicides. Sulfonylurea herbicides exist that show selectivity to numerous commercially important crops. The selectivities are accounted for by differential metabolism of the sulfonylurea between the crop and the weeds [9, 3]. Metabolism of the sulfonylurea herbicides has been shown to be accomplished by a number of mechanisms [1, 9]. Similarly, the imidazolinone herbicides are subject to a number of metabolic events including oxidation, conjugation, and hydrolysis [9]. Studies of the triazolopyrimidines have shown two major metabolic mechanisms for crop tolerance. Alkyl oxidation coupled to glucose conjugation was the important mode of metabolism in several monocots and dicots; ring cleavage was important in soybean tolerance [10]. Hodges et al. (1990) concluded that since the pathway of oxidative metabolism was common to all plants tested, the pathway alone could not explain crop tolerance. Indeed, the basis of differing susceptibilities was the rate of metabolism through a common pathway [11].

Resistance in cell culture has been achieved through selection with all three families of ALS inhibitors [12, 13, 14]. An altered ALS was the most common basis of resistance. Further, sulfonylurea and imidazolinone herbicides have produced resistance in weed populations after extended use in the field [7, 15]. There are examples of cross-resistance between and within chemical families when representatives of any one of the three families were used as selection agents [7, 8, 12, 16]. The introduction of new herbicides sharing the ALS inhibition mode-of-action, coupled to the widescale use of several existing products with this mode-of action, suggests the problem of resistance and cross-resistance will continue to escalate. This expectation presents an immediate need in the field to distinguish between weeds which are resistant due to a target-site alteration and ones which did not receive a lethal dose of herbicide during application.

In this review we will summarize the most recent information on metabolism-based and target-site-based resistance to TP herbicides. We will also introduce a quick method for the determination of target-site-based resistance that could be utilized in the field.

2 Metabolism Based Resistance to Triazolopyrimidine Sulfonanilides

2.1 Soybean Selectivity

Soybean metabolism has been studied for four TP herbicides. The structures of these four compounds are illustrated in Fig. 1. Previous research has shown that soybeans are sensitive to the sulfonanilide *N*-(2,6-dichlorophenyl)-5,7-dimethyl-1,2,4-triazolo[1,5-*a*]pyrimidine-2-sulfonamide (**1**, Fig. 1), even though metabolism by alkyl hydroxylation and glucose conjugation does occur [11]. This indicated that the rate of metabolism observed was insufficient to impart a significant degree of selectivity (Fig. 1). In contrast, flumetsulam [11] demonstrates a high degree of soybean selectivity imparted by its rapid detoxification [10, 17]. The metabolism of two additional compounds with soybean selectivity is illustrated in Figs. 2 and 3.

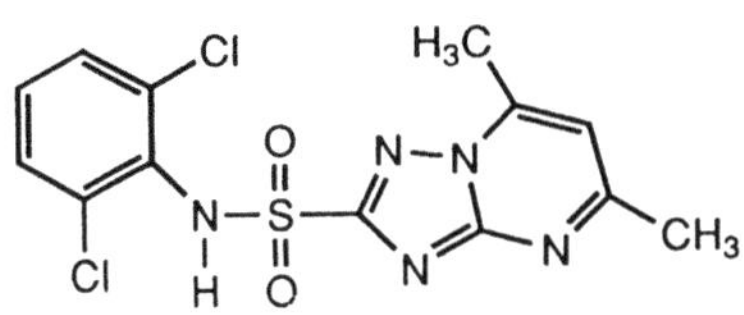

1

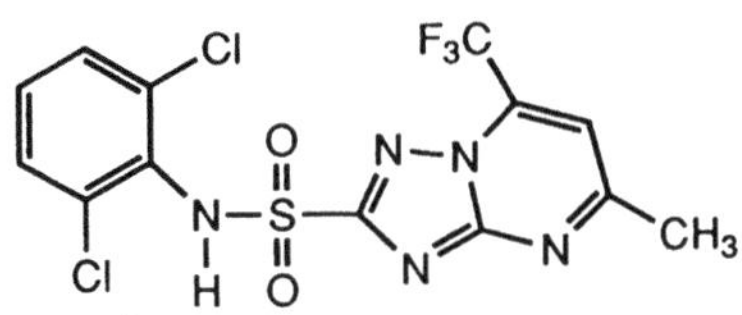

2

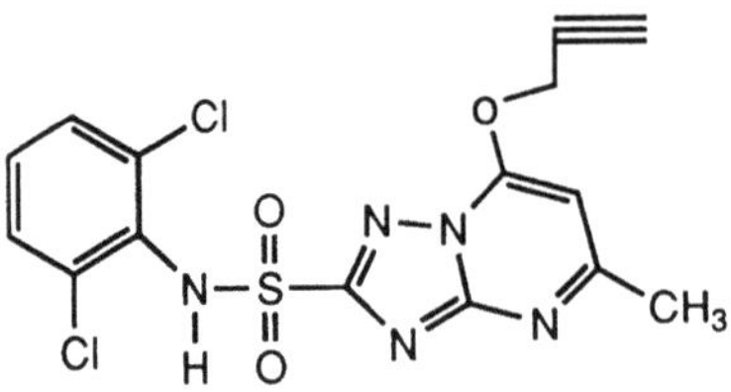

3

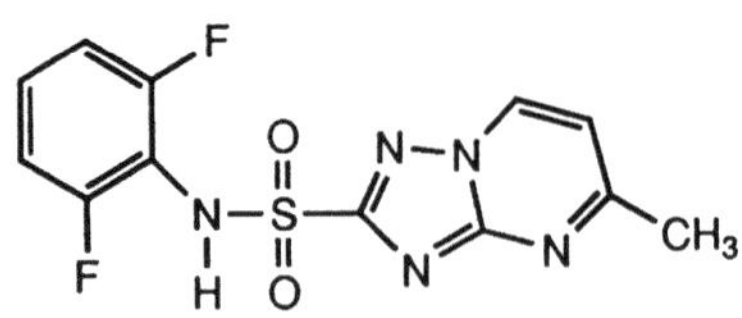

Flumetsulam

Fig. 1. Triazolopyrimidine sulfonanilides subject to metabolic inactivation in soybeans

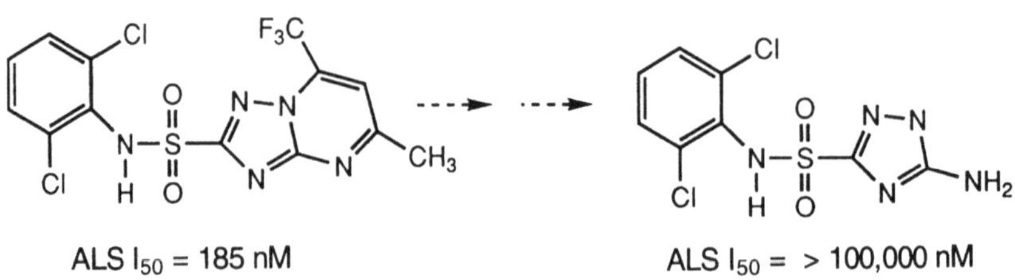

Fig. 2. Metabolism of **2** in soybeans. Analytical methods were adapted from [11]

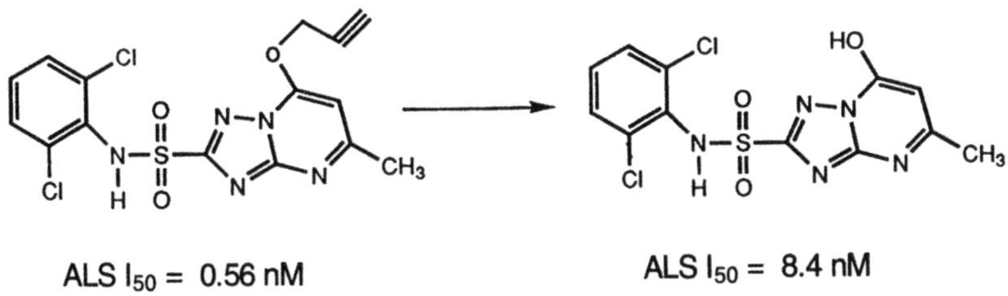

Fig. 3. Metabolism of **3** in soybeans. Analytical methods were adapted from [11]

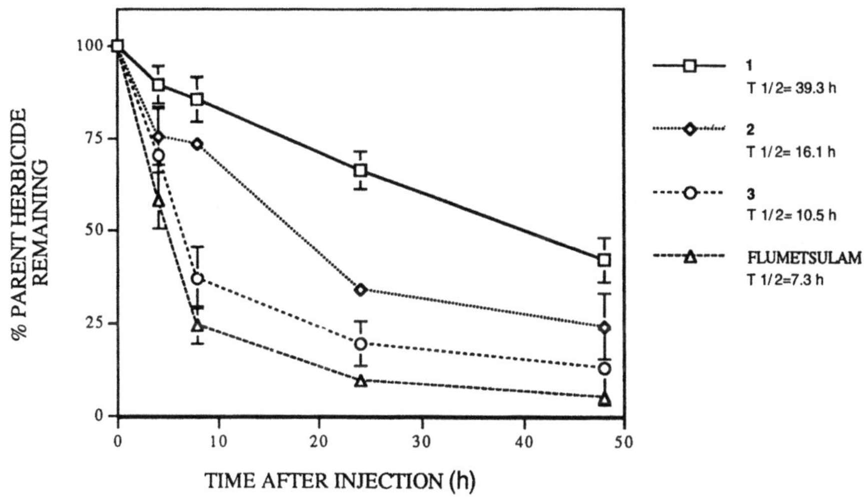

Fig. 4. Soybean metabolism of triazolopyrimidine sulfonanilide herbicides. Analytical methods were adapted from [11]

In soybean plants the primary metabolite of **2** was found to be the aminotriazole metabolite (Fig. 2). This metabolite is the product of ring cleavage. A number of the complex intermediates leading to this end product have not been identified. Based on the I_{50} for ALS, the aminotriazole metabolite is at least 500-fold less inhibitory. Ring cleavage has been shown to be important in the selectivity of other herbicides [2] and has recently been implicated in the selectivity of flumetsulam in soybeans [10, 17]. Soybeans appear uniquely able to metabolize flumetsulam by this means [10, 17].

The major metabolite of **3** was isolated from soybean plants and confirmed by NMR to be the O-dealkylation product (Fig. 3). This metabolite has an I_{50} for ALS that is 15-fold higher than the parent herbicide, making this also a highly effective detoxification mechanism. To our knowledge, O-dealkylation has not previously been reported as a mechanism of detoxification in soybeans.

However, O-dealkylation of a methyl group is well known in the metabolism of chlomethoxynil [2] and bensulfuron-methyl [18] in rice plants.

A comparison of the rate of metabolism of the above compounds with flumetsulam and Compound **1** is provided in Fig. 4. Using first order transformation, half-lives were calculated for each of the four compounds. Soybean selective compounds **2** and **3** and flumetsulam demonstrated half-lives of less than 17 h. Consistent with previous reports [11], **1** had a half-life in excess of 37 h. The inverse relationship between half-life and soybean selectivity in TP herbicides is consistent with previous reports of metabolism based selectivity [1, 9, 19].

2.2 Wheat Selectivity

Previous investigations have described the oxidation of the heterocycle as a means of detoxification in wheat for **1** [11]. In contrast to this earlier work, two examples of wheat tolerance based upon alkyl oxidation of aniline ring substituents are now known (Fig. 5). One of these compounds (proposed common name metosulam) is under commercial development for broadleaf weed control in cereals and corn [17]. Both compounds are metabolized by alkyl hydroxylation

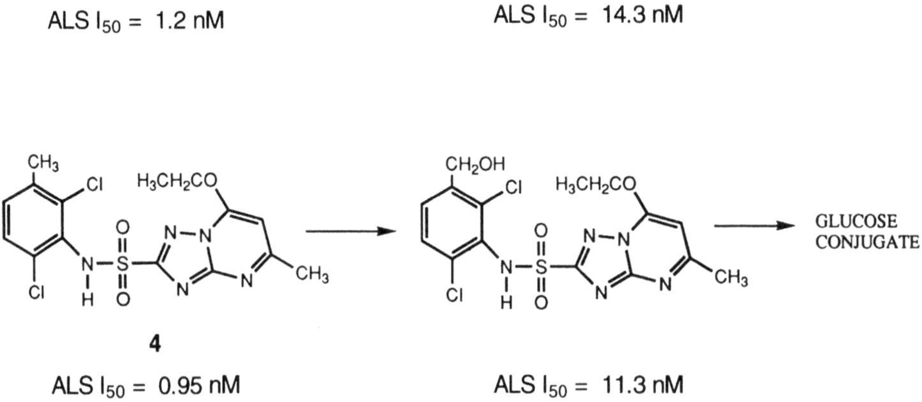

Fig. 5. Metabolism of metosulam and **4** in several plant species. Analytical methods were adapted from [11]

on the aniline ring of the molecule followed by glucose conjugation (Fig. 5). The hydroxylated metabolites of metosulam and **4** demonstrated I_{50s} for ALS that were ten times less than the parent herbicides; the glucose conjugates were completely inactive. Alkyl hydroxylation and glucose conjugation in wheat is also known to impart selectivity in SU herbicides [19].

A shared mechanism of detoxification for metosulam and **4** was also found in several broadleaf weeds. The resistance of wheat can be attributed to the differential rates of metabolism (Tables 1 and 2). These examples demonstrate that selectivity in the TP herbicide family frequently results from differential rates of metabolism between crop and weeds, and not necessarily from unique detoxification pathways afforded by the crop.

Table 1. Relationship between plant response and metabolic half-life for metosulam

Plant Type	GR_{50}[a] (ppm)	Metabolic $T_{1/2}$[b] (h)
Wheat	45	2.7
Ipomoea hederaceae	<2	75.8
Amaranthus retroflexus	<2	47.8

[a] Active ingredient postemergence application rate causing 50% reduction in growth in greenhouse tests relative to untreated controls

[b] Time required for plants to metabolize 50% of the applied compound

Table 2. Relationship between plant response and metabolic half-life for **4**

Plant Type	GR_{50}[a] (ppm)	Metabolic $T_{1/2}$[b] (h)
Wheat	6	11.7
Ipomoea hederaceae	<0.98	96
Amaranthus retroflexus	2.5	29.2
Abutilon theophrasti	2	55.6

[a] Active ingredient postemergence application rate causing 50% reduction in growth in greenhouse tests relative to untreated controls

[b] Time required for plants to metabolize 50% of the applied compound

3 Active-Site-Based Resistance to Triazolopyrimidine Sulfonanilides

3.1 Resistance in Cell Cultures

Several cell lines have been selected for resistance to TP herbicides in vitro and analyzed for cross-resistance. Subramanian et al. [20] described two cotton cell lines and one tobacco cell line resistant to TP herbicides. Previous reports included descriptions of TP resistant cell lines from soybean and tobacco as well [21]. ALS isolated from these cell lines demonstrated markedly diverse patterns of cross-resistance suggesting differing mutations in each. Cross-resistance varied from SU only, to SU and IM cross-resistance.

Substrate affinities and activities of ALS from resistant cell lines were similar to wild-type enzymes. Interestingly, a cell line that demonstrated cross resistance to SU herbicides additionally demonstrated reduced sensitivity to the feedback inhibitors valine and leucine [12]. In light of the diverse nature of mutations that can confer resistance to ALS inhibitors [22], loss of regulation appears to be a positional effect and not an indication that TP and SU herbicides occupy the regulatory site of ALS. However, branched-chain amino acid levels in seeds of SU resistant *Kochia* are higher than in susceptible *Kochia* biotypes [23]. SU resistance in weed biotypes may well be linked to a modification of the regulatory site of ALS.

3.2 Cross-Resistance to Chlorsulfuron Selected Weed Biotypes

No resistance has been selected for in whole plants to TP herbicides either through field applications or laboratory selection. Accordingly, target-site based resistance to TP herbicides in whole plants is limited to cross-resistance studies where either SU or IM herbicides served as selection agents. Although there are several reports of cross-resistance to TP herbicides in chlorsulfuron resistant biotypes [24, 8], the majority of these studies have utilized a single TP herbicide. A survey of cross-resistance across TP compounds was conducted and summarized by Subramanian et al. [20]. The results of this survey on chlorsulfuron resistant *Kochia* and *Stellaria* biotypes is provided in Tables 3 and 4.

Kochia and *Stellaria* biotypes were tested for cross-resistance both at the whole plant level and enzyme level. Six TP herbicides of diverse chemical nature, three SU herbicides, and one IM were also included for comparison. As target-site-based resistance of ALS has been reported as a semi-dominant or dominant trait [25, 15], the absolute magnitude of resistance could vary according to the heterozygosity of the seed. Accordingly, a number of replicates were utilized in the whole plant studies and ALS was extracted from a large number of plants and frozen at $-70\,°C$ for ALS sensitivity determination. The data provided in

Table 3. Comparison of whole plant and ALS sensitivity of chlorsulfuron resistant (R) and sensitive (S) *Kochia scoparia* biotypes. Adapted from [20]

Compound	Inhibitor Class	Whole Plant			ALS		
		R biotypes GR_{50} $g\,ha^{-1}$	S biotype GR_{50} $g\,ha^{-1}$	Ratio R/S[a]	R biotype I_{50} ppm	S biotype I_{50} ppm	Ratio R/S[a]
imazethapyr	IM	1.4	0.6	2.3	1.93	2.61	0.74
nicosulfuron	SU	49.9	5.2	9.5	0.35	0.12	3.0
flumetsulam-TP 1	TP	24.6	0.5	50.5	0.75	0.13	5.88
chlorimuron	SU	30.9	0.6	50.5	NA[b]	NA	NA
TP 3	TP	78.8	1.5	52.1	1.3	0.09	14.0
TP 4	TP	26.4	0.3	92.7	NA	NA	NA
chlorsulfuron	SU	79.6	0.7	114.5	0.32	0.005	63.3
TP 2	TP	138.9	0.6	241.9	0.035	0.0014	25.2
TP 5	TP	1044.8	4.0	258.4	13.0	0.7	19.4
TP 6	TP	108.9	0.3	428.6	NA	NA	NA

[a] R/S, the ratio of activity observed in resistant biotype to susceptible biotype
[b] NA, not available

Table 4. Comparison of whole plant and ALS sensitivity of chlorsulfuron resistant (R) and sensitive (S) *Stellaria media* biotypes. Adapted from [20]

Compound	Inhibitor Class	Whole Plant			ALS		
		R biotypes GR_{50} g ha^{-1}	S biotype GR_{50} g ha^{-1}	Ratio R/S[a]	R biotype I_{50} ppm	S biotype I_{50} ppm	Ratio R/S[a]
imazethapyr	IM	1.6	1.5	1.1	1.0	0.48	2.1
nicosulfuron	SU	19.1	12.0	1.6	0.17	0.07	2.4
flumetsulam-TP 1	TP	2.9	0.6	5.2	0.1	0.03	3.2
TP 4	TP	1.9	0.04	44.7	0.006	0.0007	8.1
TP 2	TP	1.5	0.03	45.3	0.006	0.0003	21.8
TP 6	TP	80.0	1.0	80.2	0.11	0.033	3.3
chlorimuron	SU	134.1	1.6	81.5	0.12	0.007	18.0
chlorsulfuron	SU	13.2	0.1	116.2	NA[b]	NA	NA
TP 5	TP	631.8	2.6	236.6	1.5	0.23	6.4
TP 3	TP	>1000	161.4	–	0.12	0.03	4.3

[a] R/S, the ratio of activity observed in resistant biotype to susceptible biotype
[b] NA, not available

Tables 3 and 4 show a relative assessment of cross-resistance between and within chemical families.

In *Kochia*, cross-resistance was observed at the whole plant level to all ALS inhibiting herbicides (Table 3). The levels varied from a resistant/susceptible (R/S) ratio of 2.3 in imazethapyr, to greater than 400 in the case of TP-6. This latter finding is noteworthy since chlorsulfuron, the original selection agent, demonstrated a cross-resistance index ratio of 114. Hence, there is clear potential for cross-resistance to TP herbicides from chlorsulfuron selected biotypes and indications that, at least in some biotypes, cross-resistance can exceed the levels of resistance observed in the original SU selection agent. While cross-resistance is high in some triazolopyrimidines, it is also quite variable. Flumetsulam, the first commercial representative of the TP herbicide family, demonstrated cross-resistance one-eighth that of TP-6. Similarly, the three SU herbicides included in the study varied by an order of magnitude in R/S ratios. The single representative of IM included in the study demonstrated the least cross-resistance.

In *Stellaria*, the cross-resistance ratio of TP-5 exceeded the value obtained with chlorsulfuron, the selection agent in the field (Table 4). This finding is similar to the high resistance described above for some TP herbicides of chlorsulfuron-resistant *Kochia* (Table 3). However, the magnitude of cross resistance was generally lower for *Stellaria* than *Kochia*; little cross-resistance to imazethapyr, nicosulfuron, or flumetsulam was observed. Interestingly, each of these three compounds are representatives of different chemical families. Clearly, the data indicate the cross-resistance potential of chlorsulfuron selected weed biotypes varies not only across ALS inhibitor families, but within each family as well. This finding suggests there are non-overlapping binding domains among classes of ALS inhibitors, and within classes as well.

In both *Stellaria* (Table 4) and *Kochia* (Table 3) there was significant correlation between resistance at the whole plant level and ALS resistance (log GR_{50} R/S ratio vs log I_{50} R/S ratio in *Kochia*, $Pr > F = 0.005$, $r_2 = 0.75$; log GR_{50} R/S ratio vs log I_{50} ratio in *Stellaria*, $Pr > F = 0.004$, $r^2 = 0.62$). Further, there was a high correlation observed in levels of cross-resistance to the various ALS inhibitors between the two biotypes ($\log I_{50}$ *Kochia* vs $\log I_{50}$ *Stellaria*, $Pr > F = 0.004$, $r^2 = 0.78$). The correlation of $\log I_{50}$ R/S ratios in *Kochia* and *Stellaria* suggests functionally equivalent mutations in the gene encoding ALS. Recent efforts to map amino acid substitutions of genes in resistant ALS [15, 26, 27] have identified two domains, A and B, of particular significance to SU resistance. In a number of chlorsulfuron resistance *Kochia* biotypes, resistance results from a single amino acid substitution in Domain A. The point mutation results in replacement of proline with one of a number of amino acids [27]. The correlation of *Stellaria* and *Kochia* cross-resistance observed in the current study, coupled to the reported prevalence of Domain A proline replacements in *Kochia* and other resistant weed biotypes, suggests the functional equivalent mutations observed here probably result from differing amino acid substitutions in the Domain A region.

3.3 Cross-Resistance to Imazaquin Resistant Weed Biotype

Imazaquin-resistant *Xanthium strumarium* was identified in Mississippi in 1991. Acetolactate synthase isolated from these plants demonstrated a high level of resistance to imazaquin (Fig. 6). There was no cross-resistance of the resistant enzyme to flumetsulam (Fig. 1) or to chlorimuron [7]. The lack of cross-resistance observed to these representative TP and SU herbicides is consistent with the low frequency of cross-resistance observed to SU herbicides in imazaquin selected *Datura* cell lines [16]. The results of this study, coupled to the

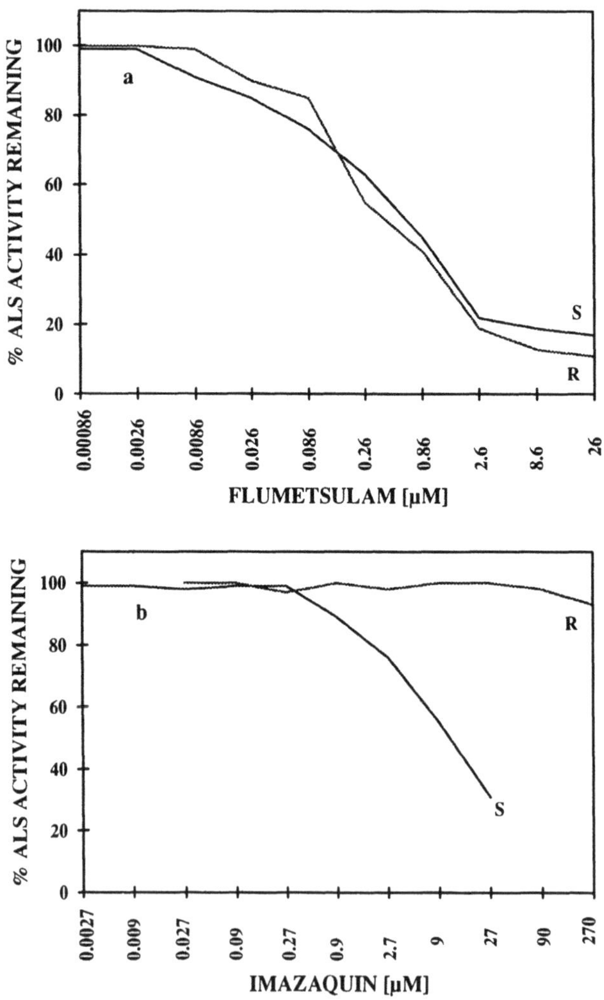

Fig. 6a, b. Inhibition of ALS isolated from susceptible (S) and imazaquin-resistant (R) *Xanthium strumarium* by **a**) flumetsulam and **b**) imazaquin. Adapted from [7]

variable cross-resistance observed to SU selected biotypes described previously, indicates that cross-resistance of IM and SU resistant weed biotypes to TP herbicides should not be assumed. However, these initial studies also suggest that cross-resistance to TP herbicides appears more likely to occur in SU selected weed biotypes, than in IM selected biotypes.

3.4 Identifying ALS Target-Site Based Resistant Plants

Identifying and preventing seed set of weeds resistant to ALS inhibiting herbicides can be part of an overall strategy aimed at preserving the usefulness of these products. To date, identification of SU and IM resistance has required sampling plant material and analyzing comparative growth rates on ALS inhibitor supplemented media. Alternatively, seeds of the plants have been collected, planted, and the sensitivity of the resulting seedlings determined. Confirmation of target-site based resistance has required the isolation and subsequent assaying of ALS from the resistant plants. While these methods are effective, they are slow and generally expensive. Most importantly, confirmation of resistance frequently comes after seed set of the putative resistant weeds.

As a part of developing an overall resistance management strategy for TP herbicides, we developed a rapid and inexpensive means to identify plants with target site based resistance [28]. The method is based upon measuring rates of acetoin accumulation in the presence and absence of an ALS inhibitor herbicide. Acetolactate accumulation is induced by inhibition of ketol-acid reductoisomerase (KARI, EC 1.1.1.86), the enzyme immediately following ALS in branched chain amino acid biosynthesis. In the presence of a KARI inhibitor, carbon flow beyond acetolactate is blocked and acetolactate rapidly accumulates. The amount of acetolactate can be quantified by homogenization, decarboxylation of acetolactate to acetoin, and colorimetric determination of the latter [29].

The influence of the KARI inhibitors 1,1-cyclopropanedicarboxylic acid (CPCA) and HOE 704 on acetoin accumulation in leaf disks is shown in Table 5. Acetolactate accumulation is rapid under these conditions and sufficient levels are present for detection by decarboxylation to acetoin and visualization with napthol and creatine in 2–12 hours [28]. In the absence of KARI inhibition no acetoin can be detected (Table 5). Similarly, if carbon flow is blocked prior to KARI, which occurs when an ALS inhibitor is present in the incubation medium, acetoin accumulation is markedly reduced (Table 5). The difference in acetoin accumulation observed in the presence and absence of an ALS inhibitor forms the basis for distinguishing between ALS sensitive and ALS resistant biotypes.

Metabolism of TP herbicides has previously been discussed as profoundly affecting the relative sensitivity of plant species. In the case of flumetsulam, aniline ring hydroxylation (*Chenopodium*), methyl group oxidation (*Amaranthus*), and ring oxidation (soybeans) have all been shown to deactivate the parent herbicide [10]. Accordingly, it was of interest to know if metabolism

Table 5. Effect of CPCA and HOE 704 alone, or in combination with flumetsulam, on acetoin accumulation in velvetleaf. Adapted from [28]

HOE-704 (μM)	CPCA[a] (μM)	Flumetsulam (μM)	Acetoin (μg gfw^{-1} h^{-1})
—	—	—	0
100	—	—	17.6
100	—	10	0.1
—	100	—	8.5
—	100	10	0.1

[a] CPCA; 1,1-cyclopropanedicarboxylic acid

Table 6. Comparison of metabolic stability, whole plant sensitivity, and inhibition of acetoin accumulation activity of flumetsulam in four different species

Plant Type	$T_{1/2}$[a] (h)	GR_{80}[a] (g ha^{-1})	I_{50} (μM)
soybean	18	>280	0.150
Chenopodium album	131	20	0.045
Amaranthus retroflexus	104	25	0.040
Abutilon theophrasti	no metabolism	22	0.090

[a] Adapted from [10]

based resistance to TP herbicides could be distinguished from active-site based resistance through the acetoin-based diagnostic assay. The sensitivity of CPCA-induced acetoin accumulation to flumetsulam was examined in four different species varying both in pathway of detoxification and half-life for detoxification. These data are summarized in Table 6 along with comparative sensitivity of these species at the whole plant level. Consistent with previous reports [11], a relationship between half-life and whole plant sensitivity is noticeable (Table 6). The sensitivity of acetoin accumulation to flumetsulam, however, appears nearly equal among all four species with no relationship to half-life. The differences in I_{50} concentrations are less than two-fold between soybeans, a highly tolerant species, and *Abutilon theophrasti*, a highly susceptible species. The acetoin-based diagnostic assay appears specific for target-site-based resistance.

The diagnostic method was validated by clearly distinguishing between imazaquin sensitive and imazaquin resistant *Xanthium strumarium* biotypes (Table 7). In the presence of imazaquin, acetoin accumulation was markedly reduced in the S biotype, but not in the R. The visual difference in color of the

Table 7. Effectiveness of imazaquin at preventing CPCA-induced acetoin accumulation in resistant (R) and sensitive (S) *Xanthium strumarium* biotypes. Adapted from [28]

Xanthium biotype	CPCA[a] (μM)	Imazaquin (μM)	Reaction color	Acetoin ($\mu g\, gfw^{-1}\, h^{-1}$)
S	500	0	Pink	2.6
S	500	10	Brown	0.7
R	500	0	Pink	1.6
R	500	10	Pink	1.7

[a] CPCA; 1,1-cyclopropanedicarboxylic acid

reaction tubes was more than sufficient to distinguish qualitatively between R and S biotypes. This finding suggests the method may have promise for development of a field diagnostic kit. Whether the analysis is completed entirely in the field, or most conveniently sampled from the field and processed in the laboratory, the rapid identification of ALS target-site resistance is an important new method to help monitor and manage resistance.

Acknowledgements. The authors would like to thank Cindy Runyon for her assistance in preparing the document for publication, and Csaba Cséke for his critical review of the manuscript.

4 References

1. Brown HM, Fuesler TP, Ray TB, Strachan SD (1991). In: Frehse H (ed) Pesticide chemistry: advances in international research, development and legislation, 1st edn. VCH, p 257
2. Hatzios KK, Penner D (1982). Metabolism of herbicides in higher plants. Burgess, Minneapolis
3. Beyer EM Jr, Duffy MJ, Hay JV, Schlueter DD (1987). In: Kearney PC, Kaufman DD (eds) Sulfonylureas. Marcel Dekker, New York, p 117
4. Shaner DL, Anderson PC, Stidham MA (1984). Plant Physiol 76: 545
5. Gerwick BC, Subramanian MV, Loney-Gallant VI (1990). Pestic Sci 29: 357
6. Schloss JV, Ciskanik LM, VanDyk D (1988). Nature 331: 360
7. Schmitzer PR, Eilers RJ, Cséke C (1993). Plant Physiol (in press)
8. Hall LM, Devine MD (1990). Plant Physiol 93: 962
9. Brown HM, Neighbors SM (1987). Pest Biochem Physiol 29: 112
10. Chang M, Brown SM, Swisher BA, DeBoer GJ, Zakett D, McKendry LH, Roth GA, Stanga MA (1992). Abstr Agro 30, 204th ACS National Meeting
11. Hodges CC, DeBoer GJ, Avalos J (1990). Pestic Sci 29: 365
12. Subramanian MV, Hung HH, Dias JM, Miner VW, Butler JH, Jachetta JJ (1990). Plant Physiol 94: 239
13. Chaleff RS, Ray TB (1984). Science 223: 1148

14. Saxena PK, King J (1988). Plant Physiol 86: 863
15. Mallory-Smith CA, Thill DC, Dial MJ, Zemetra RS (1990). Weed Technol 4: 787
16. Saxena PK, King J (1990). Plant Physiol 94: 1111
17. Swisher BA, Gerwick BC, Chang M, Miner VW, deBoer GJ (1991). Abstr WSSA 31: 50
18. Brown HM (1990). Pestic Sci 29: 263
19. Brown MA, Chiu TY, Miller P (1987) Pest Biochem Physiol 27: 24
20. Subramanian MV, Gerwick BC, Martin E, Loney-Gallant V (1991). Abstr WSSA 31
21. Subramanian MV, Loney V, Pao L (1989). In: Copping, LG, Dalziel J, Dodge AD (ed) Prospects for amino acid biosynthesis inhibitors in crop protection and pharmaceutical chemistry. Society of Chemical Industry, Great Britain, p 97
22. Guttieri MJ, Eberlein CV (1993). Abstr WSSA 33: 193
23. Dyer WE, Chee PW, Fay PK (1993). Weed Sci 41: 18
24. Devine MD, Marles MAS, Hall LM (1991). Pestic Sci 31: 273
25. Falco SC, Dumas KS (1985). Genet 109: 21
26. Wiersma PA, Schmiemann MG, Condie JA, Crosby WL, Moloney MM (1989) Mol Gen Genet 219: 413
27. Guttieri MJ, Elberlein CV, Mallory-Smith CA, Thill DC, Hoffman DL (1992). Weed Sci 40: 670
28. Gerwick BC, Mireles LC, Eilers RE. (1993). Weed Tech (in press)
29. Westerfeld WW (1945). J Biol Chem 161: 495

Salicylic Acids and Related Structures

M. W. Drewes

Bayer AG Monheim, Geb. 6550, 51368 Leverkusen, Bayerwerk, FRG

The pyrimidyloxy salicylic acids and related derivatives represent a new class of compounds which inhibit the enzyme acetolactate synthase in plants. These highly active compounds control a large number of troublesome weeds and can be applied pre- or post emergent. In this new class of herbicidal compounds there are at present three development products – KIH-2031, KIH-6127, KIH-2023.

Chemistry of Plant Protection, Vol. 10

1 Introduction

The sulphonylureas, the imidazolinones and the triazolopyrimidines all belong to the group of herbicides which inhibit the plant enzyme acetolactate synthase. Recently, a fourth member consisting of hydroxy or mercapto carboxylic acid derivatives (**1** and **2**) has been found.

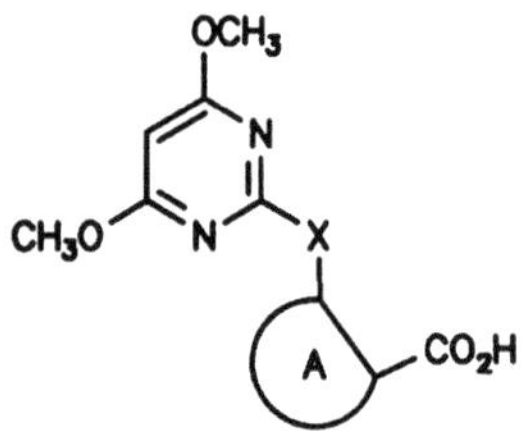

1

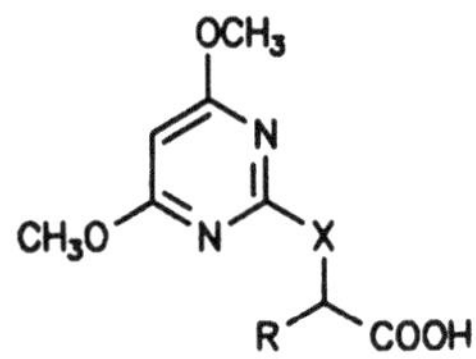

2

X=O, S
A=Aryl, Heteroaryl
Cycloalkyl

R=Alkyl, Aryl

These compounds appear to be structurally related to the sulphonylureas (Scheme 1). The terminal groups are almost identical and the only difference lies in the middle portion. Here, the sulphonylurea bridge has been reduced to a single heteroatom.

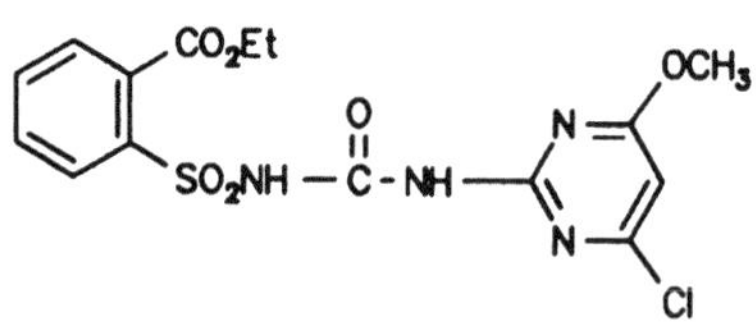

Sulphonylurea

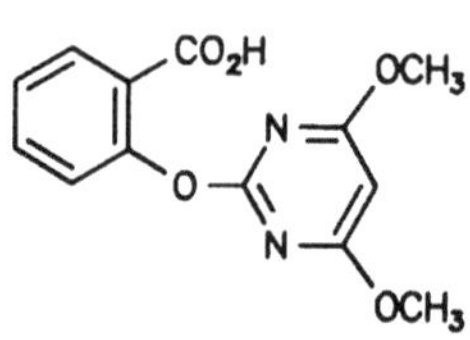

Salicylic Acid Derivative

Scheme 1

These hydroxy or mercapto carboxylic acid derivatives also control a wide spectrum of broad leafed weeds as well as numerous grassy ones. Furthermore they appear to have pre- and/or post-emergent activity.

1.1 Historical Background

In 1977 ICI [1] filed a patent claiming herbicidal properties for the phenoxy pyrimidine derivatives **3**.

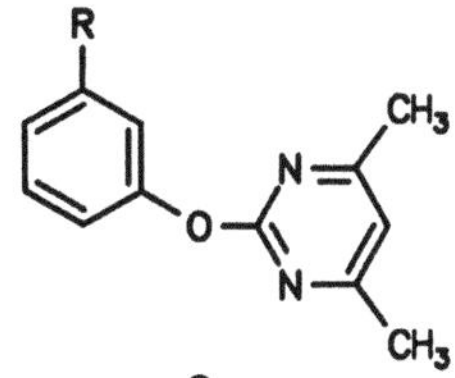

3

R=Alkyl, Alkoxy, Halogen

Almost ten years elapsed before the Kumiai Chemical Industry Company [2] synthesized the compounds with the correct substitution pattern **4**. The routes which led to the discovery of the lead compound are illustrated in Scheme 2.

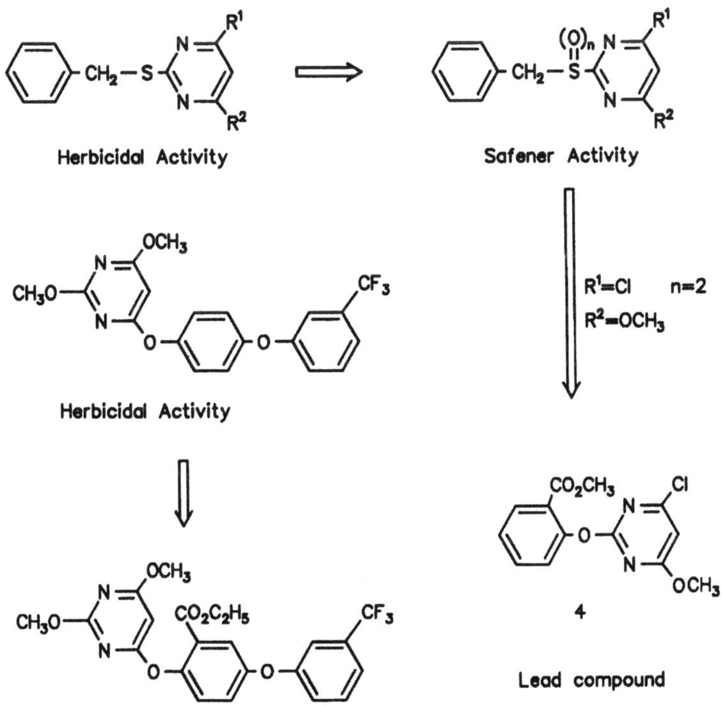

Scheme 2

Subsequently, many companies have initiated research in this field. In the last five years more than sixty patents have been filed, with the major contributions originating from Kumiai, BASF, Schering, Shell, Mitsubishi Petrochemical, Mitsui Toatsu and Bayer.

2 Pyrimidyloxy Salicylic Acid Derivatives

In order to simplify this report, it is necessary to draw attention to a number of fundamental relationships between the chemical structure and the activity [3].

2.1 Variation of W

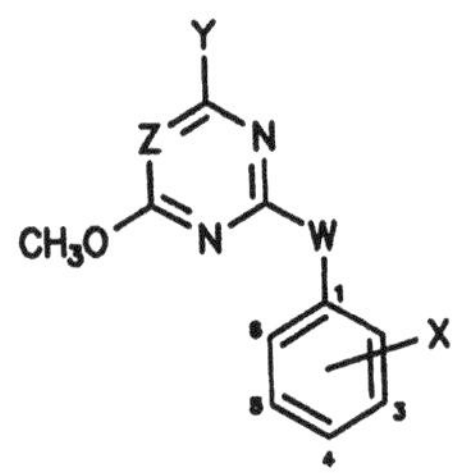

W can be either oxygen or sulphur.

2.2 Position of X

X	Herbicidal Activity
2-CO_2H	active
3-CO_2H	inactive
4-CO_2H	inactive

It is clear that the 2-position is the most favoured one. Consequently all subsequent derivatives will contain a substituent in this position.

2.3 Variation of X in the 2-position

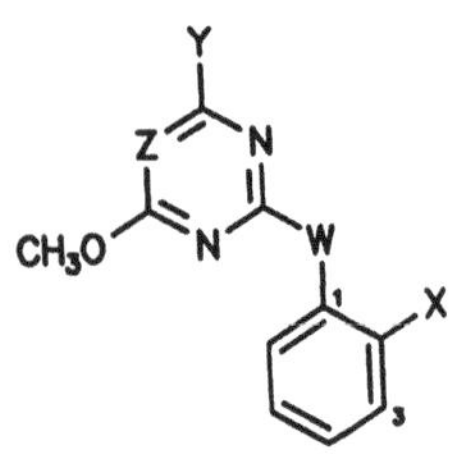

4b

X	Herbicidal Activity
CO_2H, CO_2CH_3 CHO, CH_2OH	active
H, CN, NO_2, OR CF_3, SO_2NR_2, NR_2	inactive

The herbicidal activity appears to be limited to carboxylic acid and aldehydic derivatives.

2.4 Variation of Y

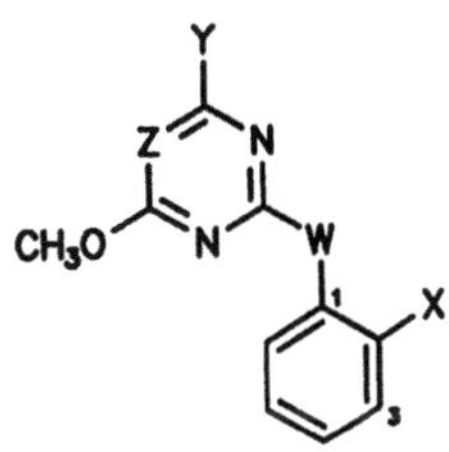

Y	Herbicidal Activity
H, Cl, CH_3, OCH_3, OEt, $OCHF_2$, SCH_3, $N(CH_3)_2$	active
Et, CF_3, Ph, OPh	inactive

The derivative with Y = OCH_3 is the most active pyrimidine and will be used as the reference heterocycle throughout this report.

2.5 Variation of Z

Z	Herbicidal Activity: (4 kg/ha)
CH	strong
N	weak

The pyrimidyl derivatives appear to be more active than the corresponding triazinyl ones. Consequently all future discussions will revolve around the pyrimidyl group.

2.6 Summary

From the previous chapters, it is possible to depict the "ideal" molecules **5** with all the "best" groups in the required positions.

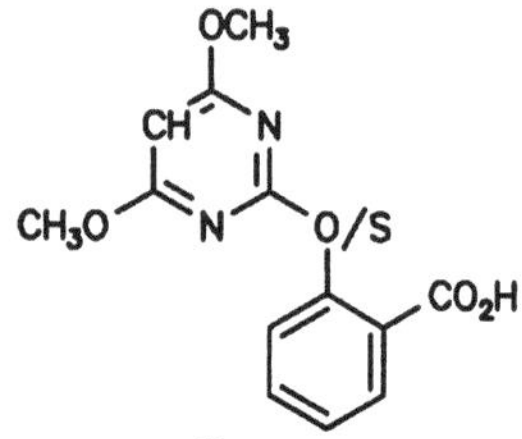

5

3 Chemical Structure Variations

Since 1987, the "lead" structure has been adapted and modified to produce a whole host of new analogues. For a classification of these modifications, it is necessary to discuss the variables X, Y and Z as in **6**.

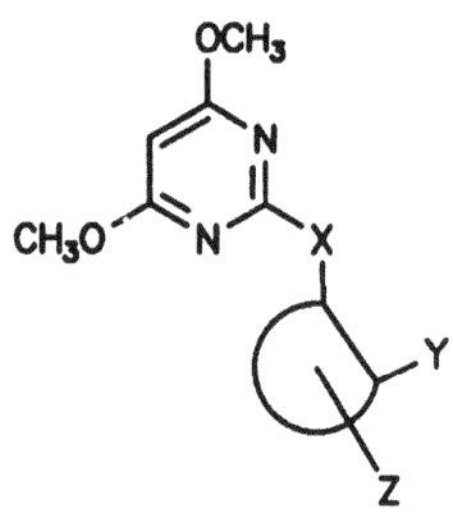

6

3.1 Variation of X

For the majority of patent applications, the variable X is either oxygen or sulphur. In some cases, X can also be NH or a substituted amino group. The oxygen and sulphur analogues show considerable herbicidal activity and from recent data [3] the following structure-activity relationships can be postulated.

1) For the activity against grasses,

$$O \gg S$$

2) for the activity against broad-leafed weeds,

$$O > S$$

3) for the safety in cotton

$$S \gg O$$

The latter is substantiated by the experimental product KIH-2031 (**7**) (see Sect. 7).

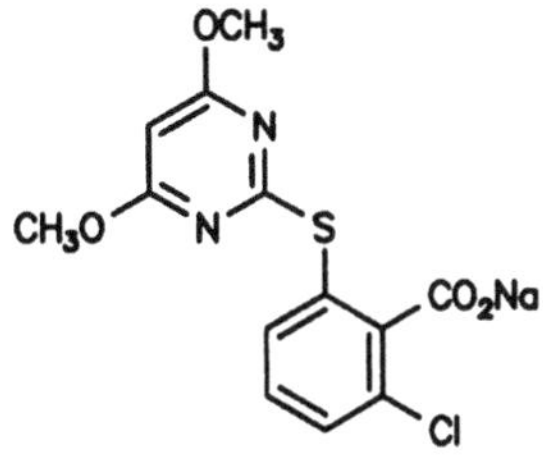

KIH-2031
7

3.2 Variation of Y on the Benzene Ring

The variation of the substituent Y in the aromatic ring has produced a host of compounds in which either the carboxylic acid or the aldehyde moiety has been functionalized.

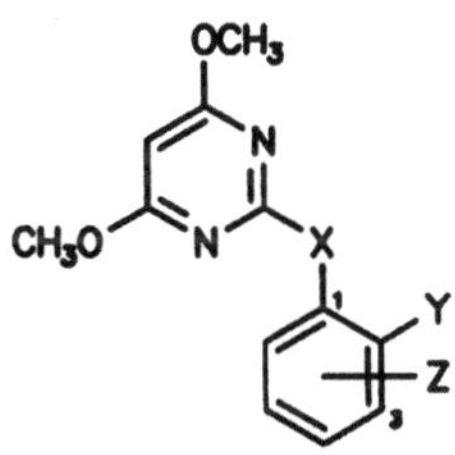

Generally, these products are used for the first biological screening in a new "prototype" and at the same time serve as starting materials for other transformations. The creative talents of organic chemists have produced a large number of ester and thioester analogues [2, 4–6]. Some examples are shown in Scheme 3, pp. 168, 169.

The synthesis of one such derivative [2] is illustrated in Scheme 4, p. 170.

Recently, a large number of amide or sulfonyl amide derivatives have been patented. Examples of such cyclic amides, amino acid derivatives as well as sulfonyl amides [7–9] are shown below (Scheme 5, p. 171).

3.2.1 The Aldehyde Moiety

The effort required to manipulate the acid functionality is no longer a trivial matter. Consequently other patents have been filed where the aldehyde moiety has been derivatized (Scheme 6, p. 171).

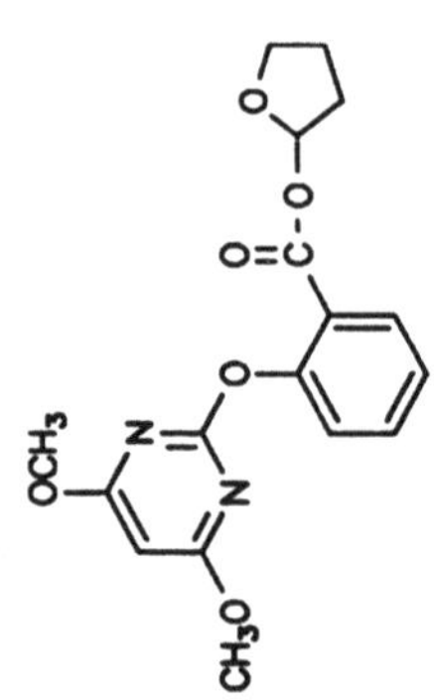

EP 457505 [5]

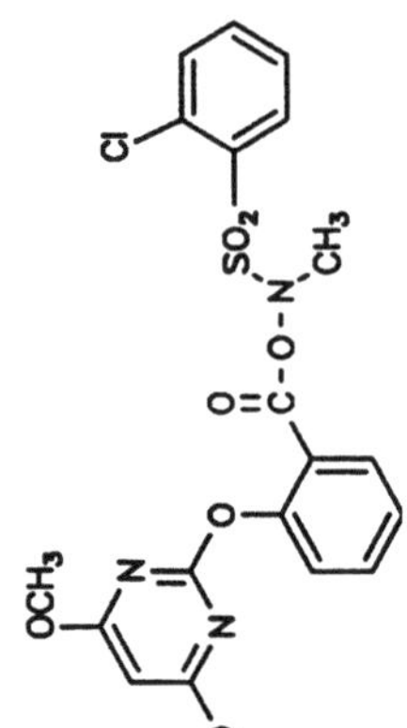

EP 468690 [6]

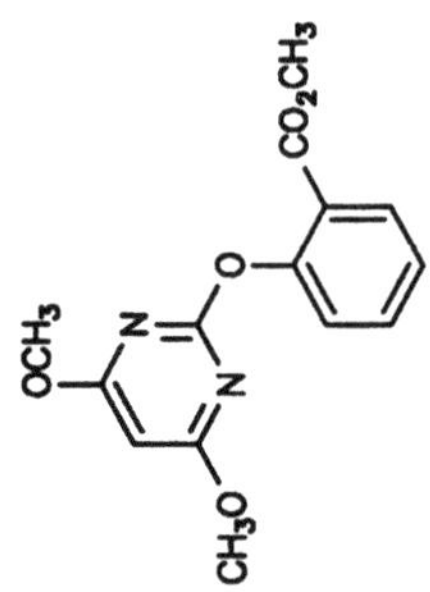

EP 223406 [2]

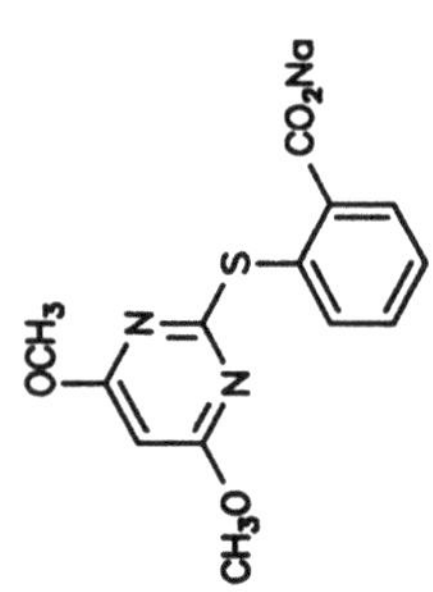

EP 223406 [2]

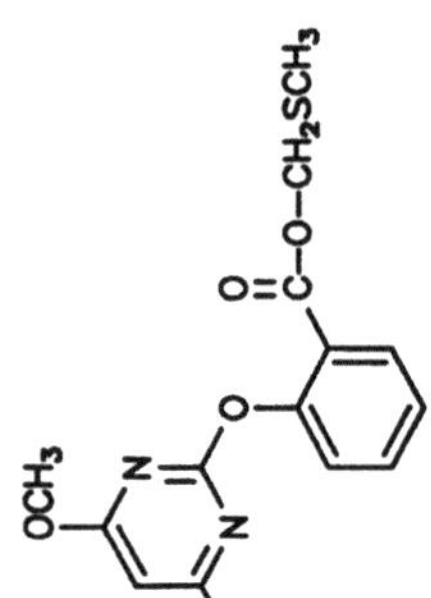

EP 223406 [2]

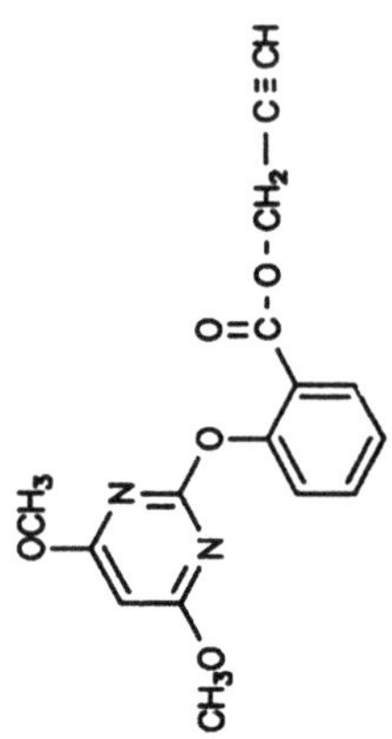

EP 223406 [2]

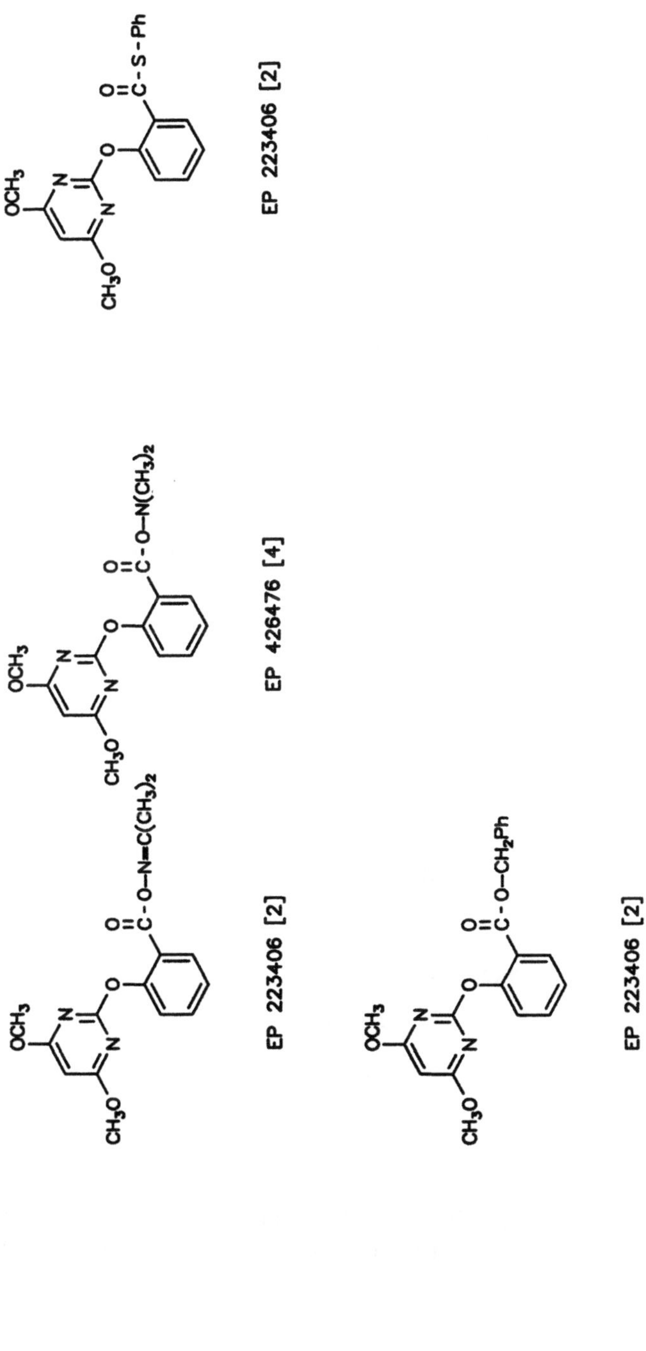

Scheme 3

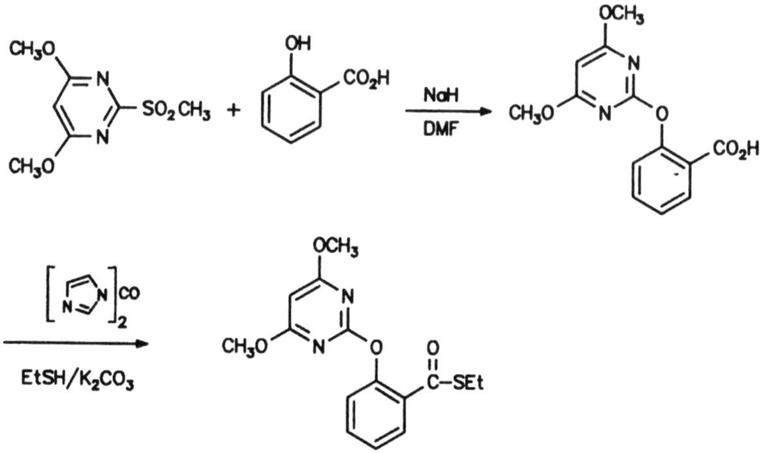

Scheme 4

Some patented examples [10–15] of such alcohol, hydrazone, oxime, acetal and α–β unsaturated compounds are shown in Scheme 7, p. 172.
Whether these derivatives have any advantages over the carboxylic acid one is at present unknown. They may serve as a transport form to allow permeation through the leaf cuticle and may subsequently be converted to the acid by the plant. Quantitative investigations to support such a hypothesis have not been published to date.

3.3 Variation of Z

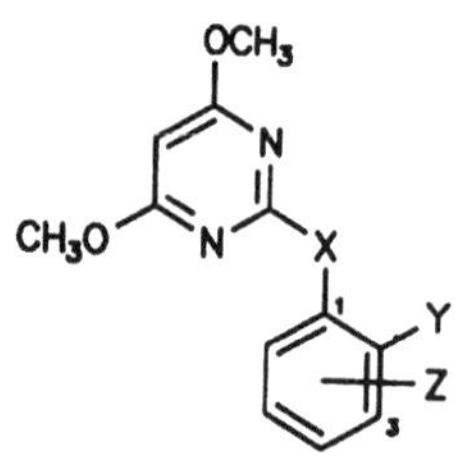

The position and nature of the substituent Z has been the subject of many variations. In most cases Z has been an alkyl, alkoxy or halogen group [16–20].

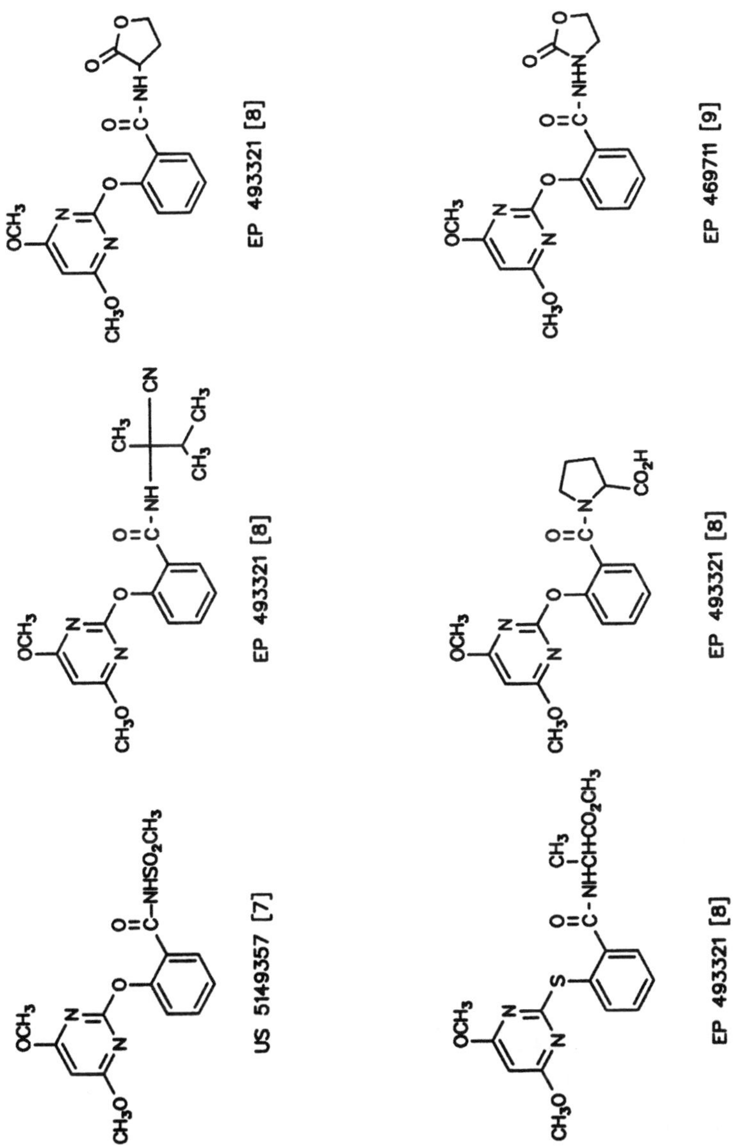
US 5149357 [7]
EP 493321 [8]
EP 493321 [8]
EP 493321 [8]
EP 493321 [8]
EP 469711 [9]

Scheme 5

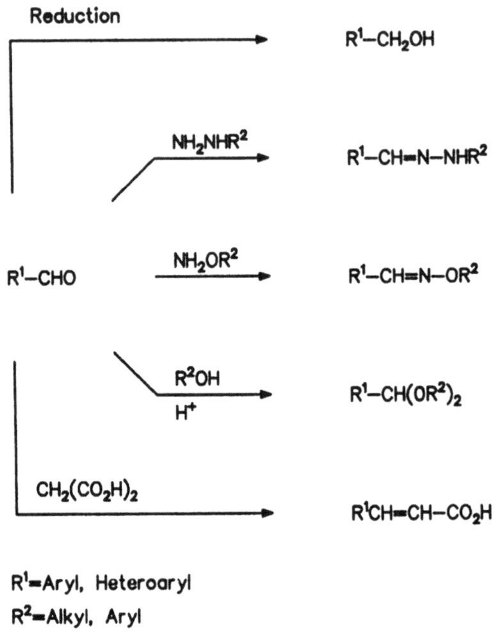

Scheme 6

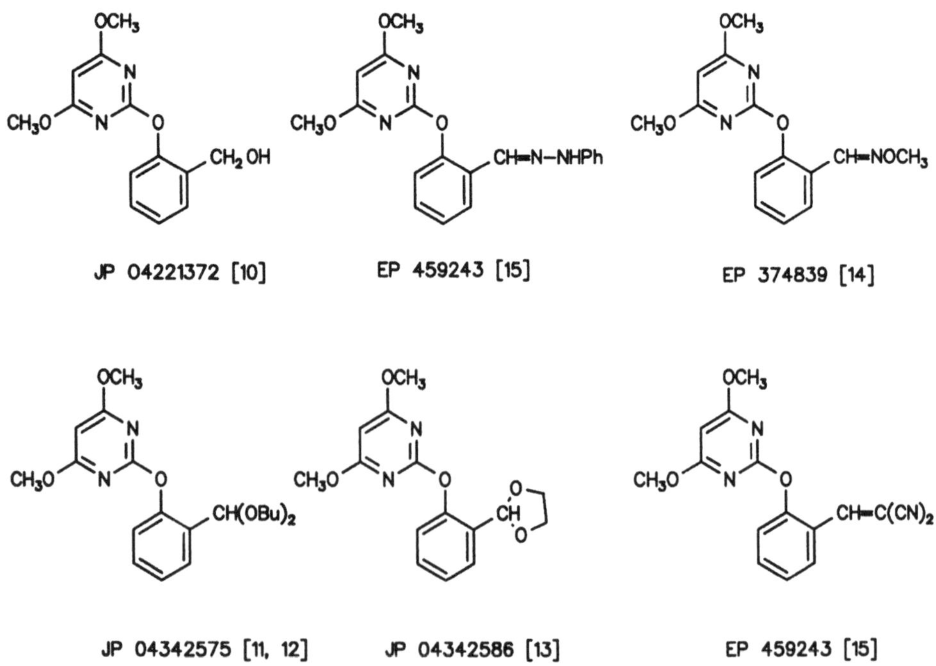

Scheme 7

At the same time, the optimal position of Z was also determined. The only published structure–activity data [3] is shown in Table 1.

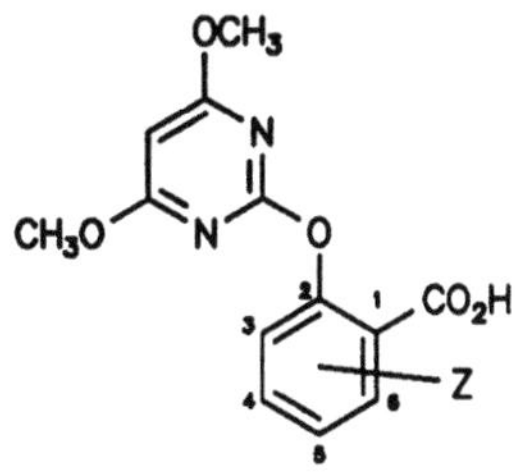

Table 1. The influence of the substitution pattern upon the pI_{50} value

Z	pI_{50}^{a}	Herbicidal Activity
H	6.6	strong
3-Cl	6.3	medium
4-Cl	4.0	weak
5-Cl	5.4	medium
6-Cl	7.6	strong

[a] Concentration that causes 50% in vitro inhibition of the acetolactate synthase

A substituent in the 6-position appears to have a positive influence on the herbicidal activity. Consequently, new variations of Z invariably involve a substituent at the 6-position [19–34]. The extent to which the Z substituent has been varied is illustrated in Scheme 8, p. 174.
The majority of these prototypes involve the synthesis of vicinal tri-substituted phenyl rings. An example [27] is seen in Scheme 9, p. 175.

3.4 Variation of the Aromatic Ring (A)

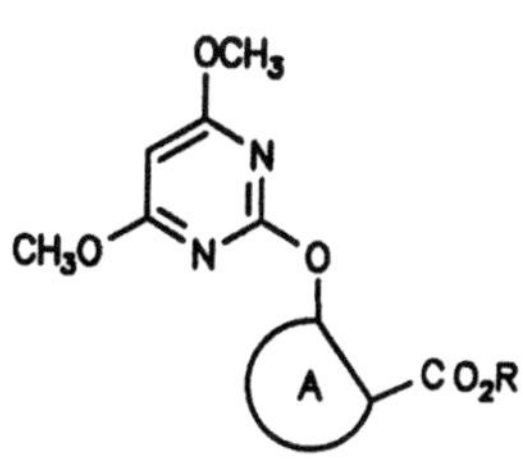

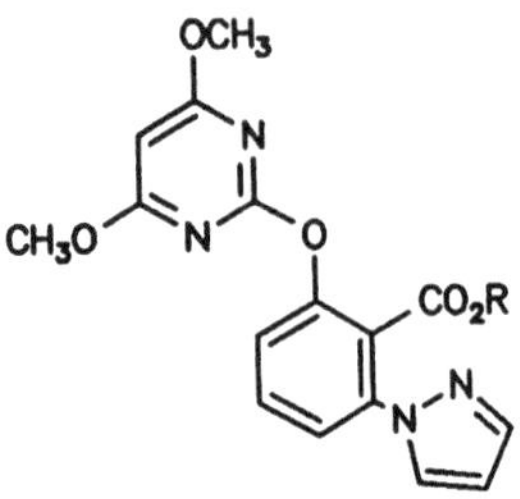

EP 402751 [27]

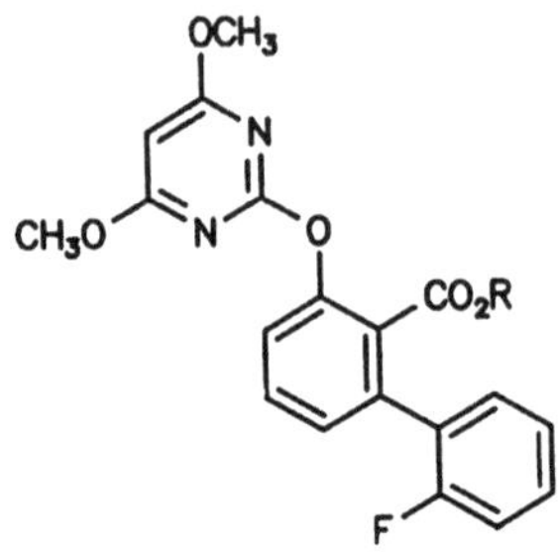

EP 402751 [27]

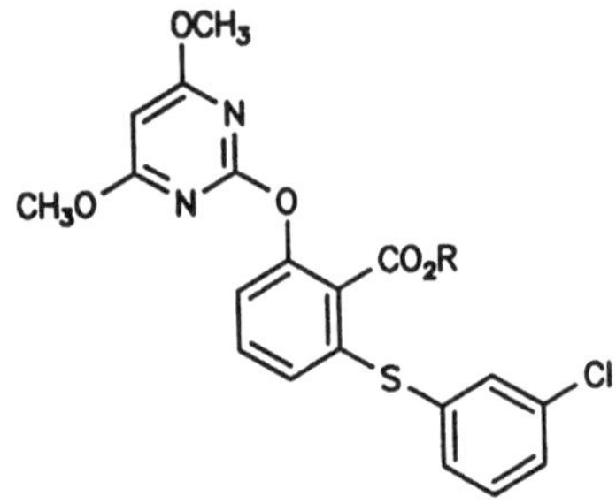

EP 435104 [29]

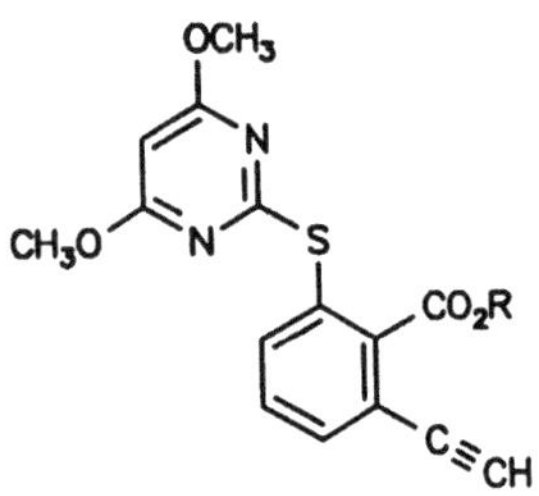

EP 490060 [30]

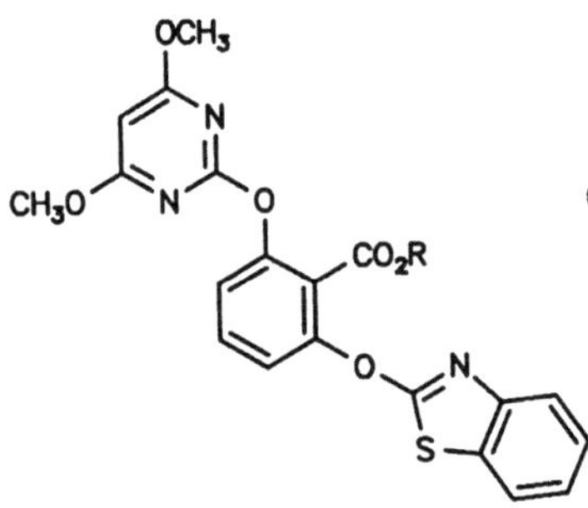

JP 0532639 [22]

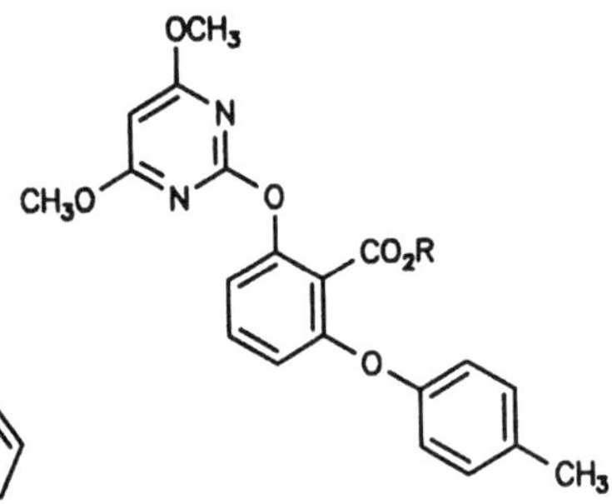

EP 287072 [19]

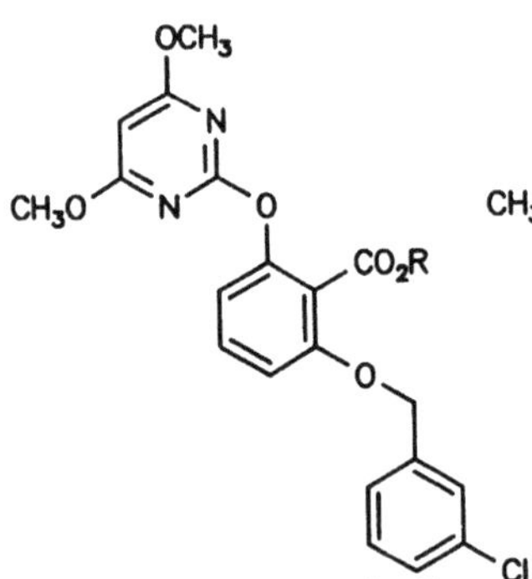

EP 287079 [20]

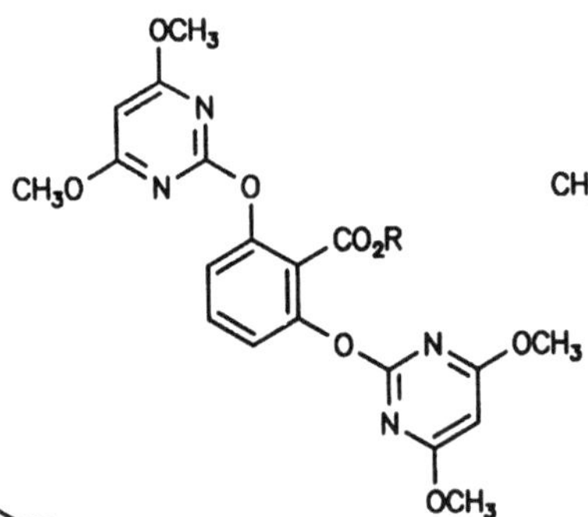

EP 321846 [21]

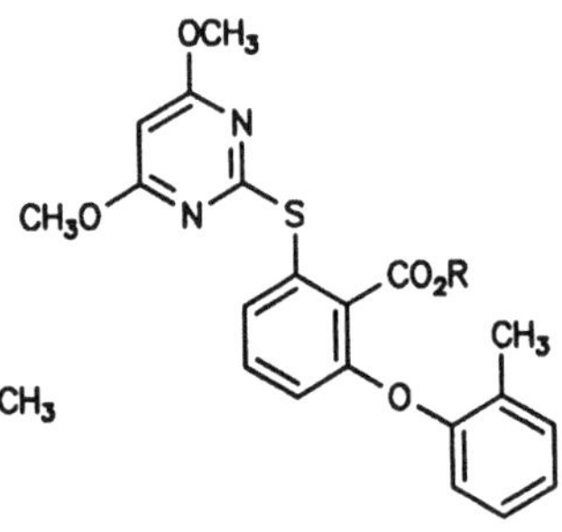

EP 315889 [17]

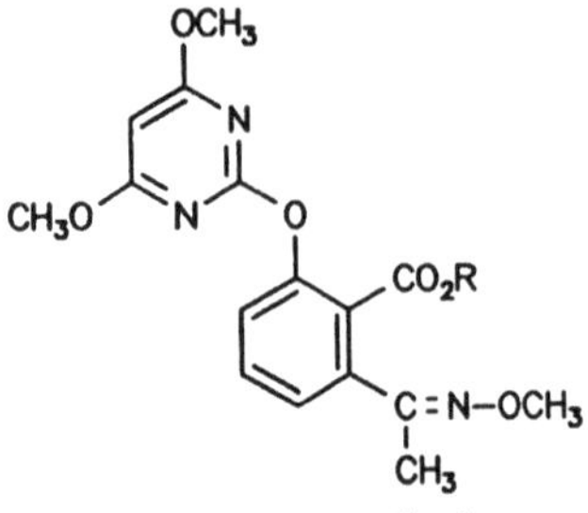

EP 435170 [25]

R=Alkyl, H

Scheme 8

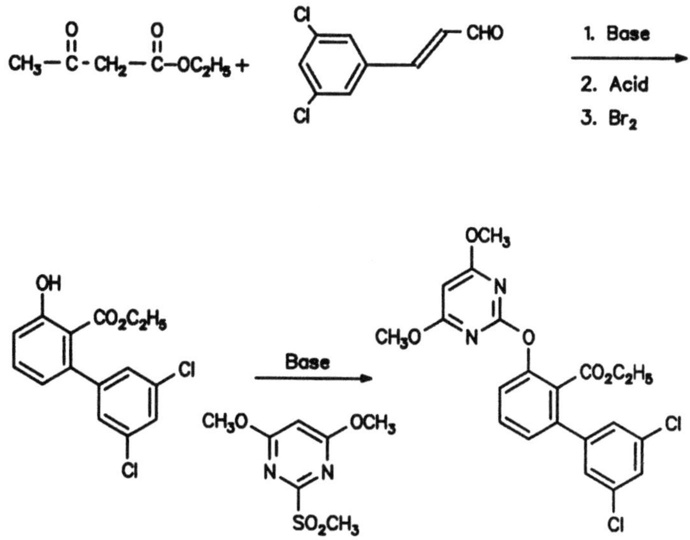

Scheme 9

After having varied the nature and the position of the substituent Z, it was only a matter of time before the benzene ring (A) was exchanged against a heterocyclic aromatic ring.
Recent patent literature [35–46] confirms this trend. Some new developments are illustrated in Scheme 10, p. 176.

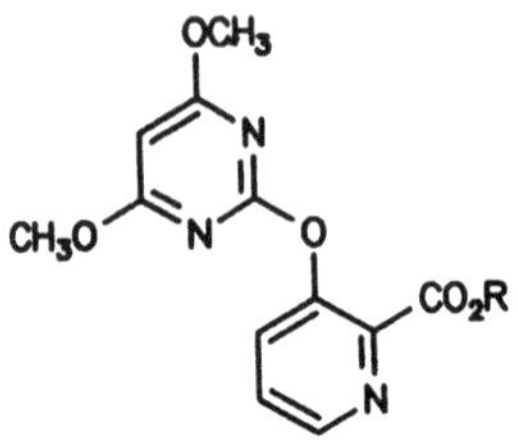

EP 249707 [35]

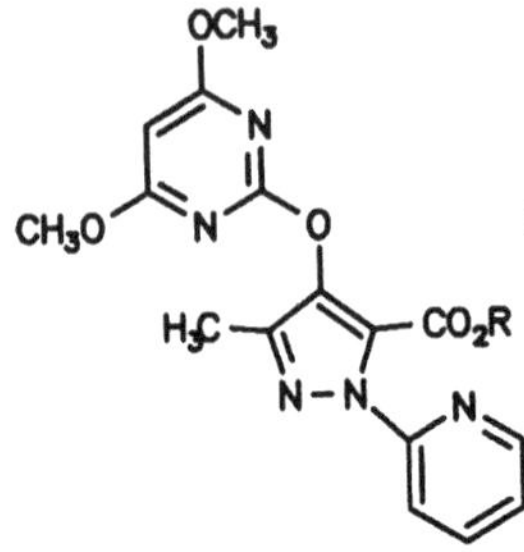

JP 04145081 [43]

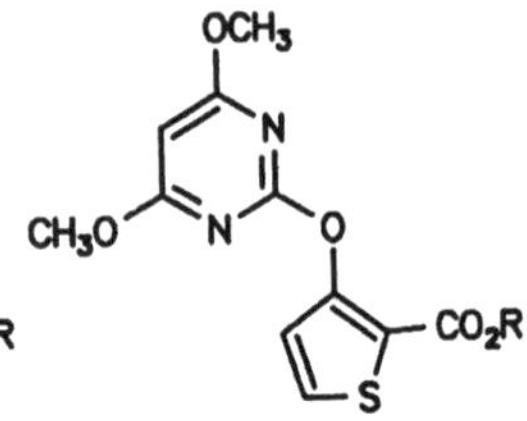

EP 360163 [39]

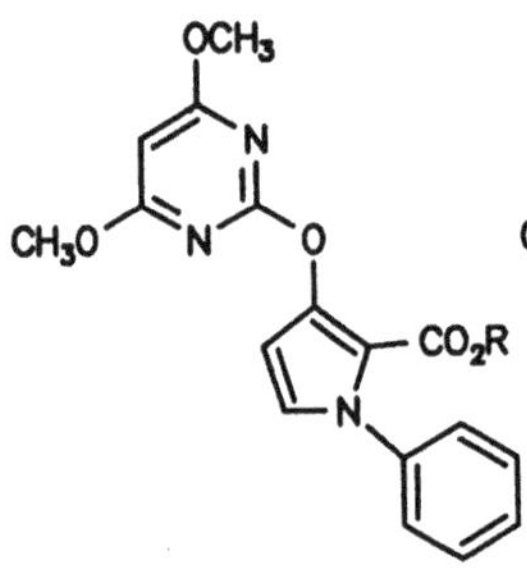

JP 03284676 [44]

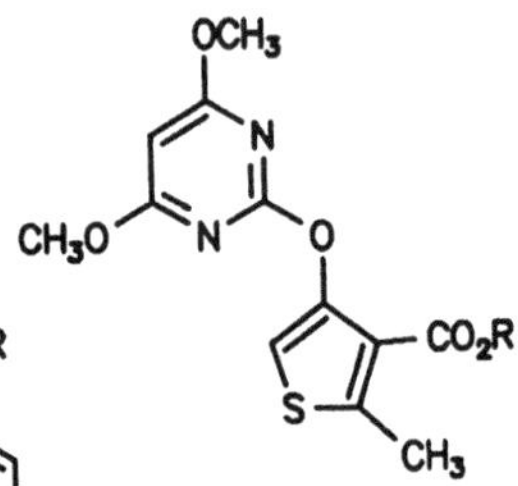

JP 03232884 [45]

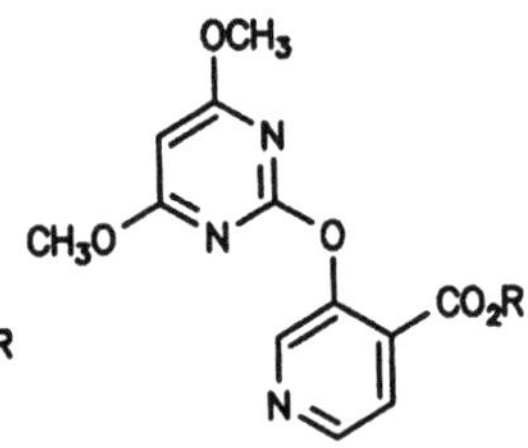

EP 360163 [39]

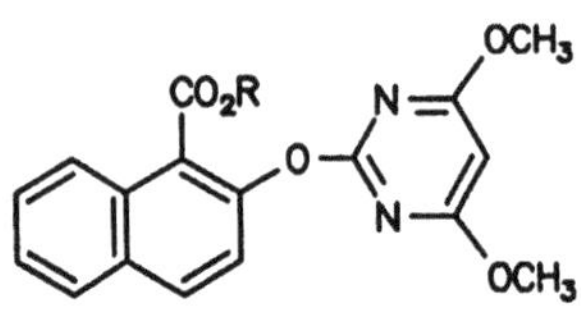

JP 02056469 [46]

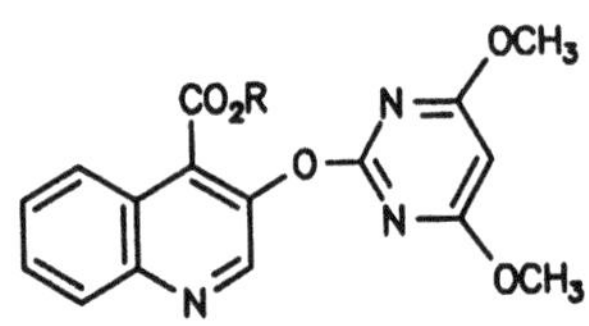

EP 360163 [39]

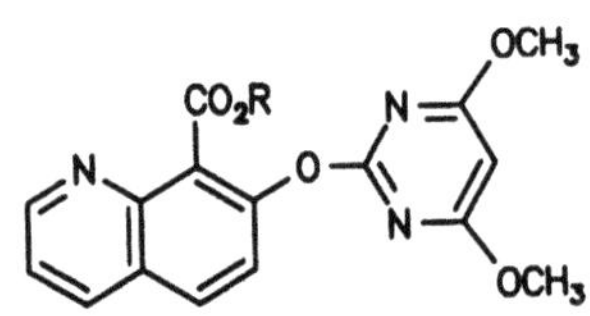

EP 360163 [39]

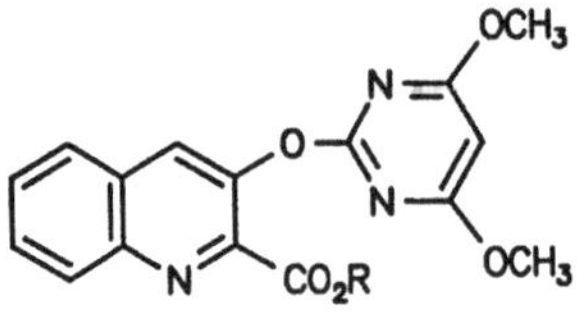

EP 360163 [39]

R=Alkyl, H

Scheme 10

The above structures show biological activity. However, there is not sufficient published data to make any meaningful structure–activity associations.

4 Acyclic and Cyclic Hydroxy/Mercapto Carboxylic Acid Derivatives

The initial research involved the manipulation of the benzene ring to synthesize new patent free compounds. Recently, a new trend has been observed: cyclic and acyclic hydroxy/mercapto carboxylic acid derivatives have been introduced to replace the benzene ring.

4.1 Cyclic Hydroxy/Mercapto Carboxylic Acid Derivatives

A number of patents [47–50] have been filed in this area of research. The majority of them contain a 5-, 6- or 7-membered ring. Some new developments are shown in Scheme 11.

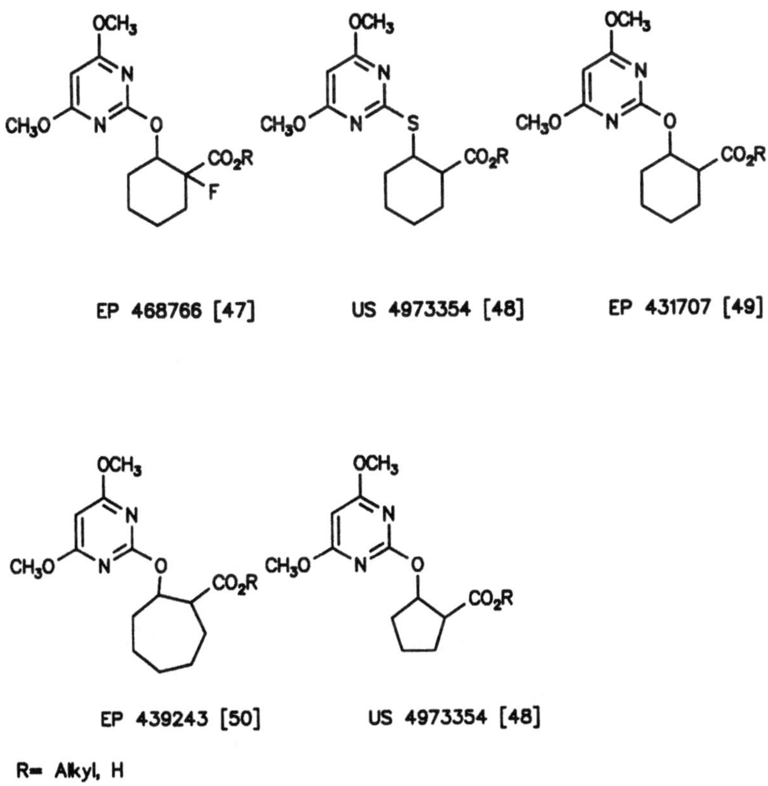

Scheme 11

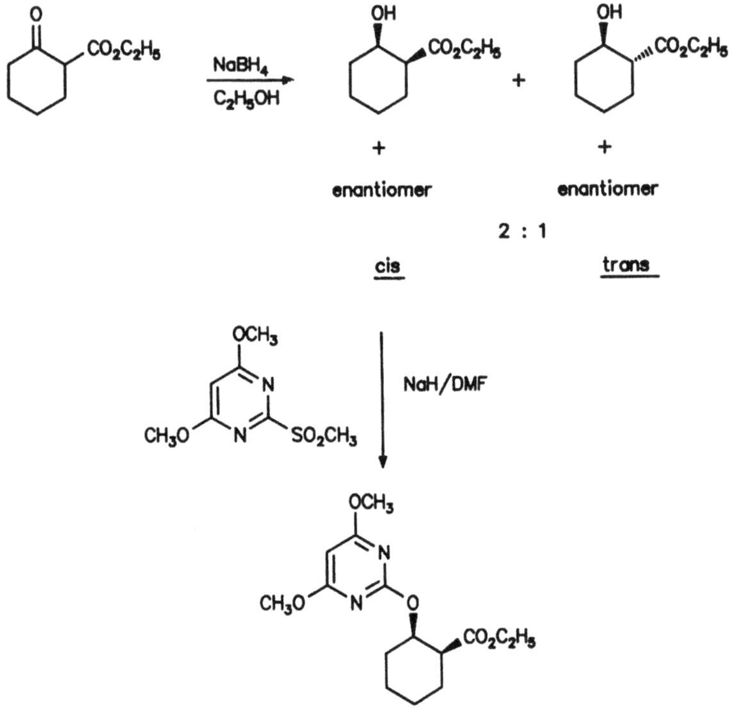

Scheme 12

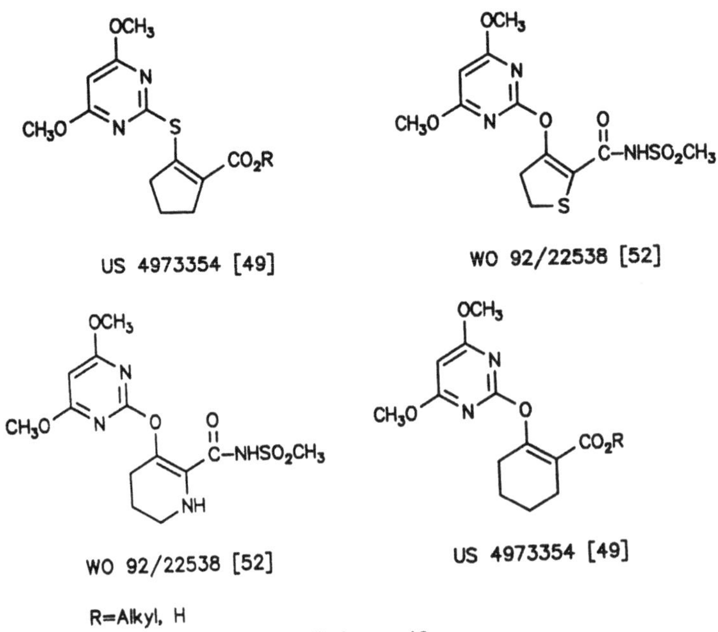

Scheme 13

These compounds are usually prepared as a mixture of diastereomers. In some cases, the isomers have been separated into their constituent diastereomers. At present, it is not known whether the various isomers have different herbicidal activity. The synthesis of a cyclic acid derivative [49] is illustrated in Scheme 12. Complementary to the saturated cyclic compounds are the unsaturated cyclic derivatives [48, 51, 52]. They are usually 5- or 6- membered rings and may contain a heteroatom. Some enol–ether/thioether derivatives are depicted in Scheme 13.
The synthesis of such compounds [49] is achieved by the following route (Scheme 14):

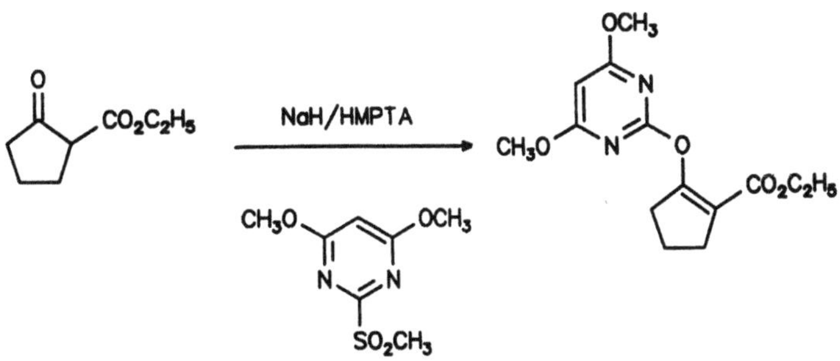

Scheme 14

4.2 Acyclic Hydroxy/Mercapto Carboxylic Acid Derivatives

The activity in this area of research has been increasing steadily. Consequently, a large number of new patents [53–63] have been filed. In the previous chapter, compounds having the hydroxy/mercapto group and the carboxylic acid vicinal to each other were described. However in the acyclic derivatives these groups are geminal to one another.

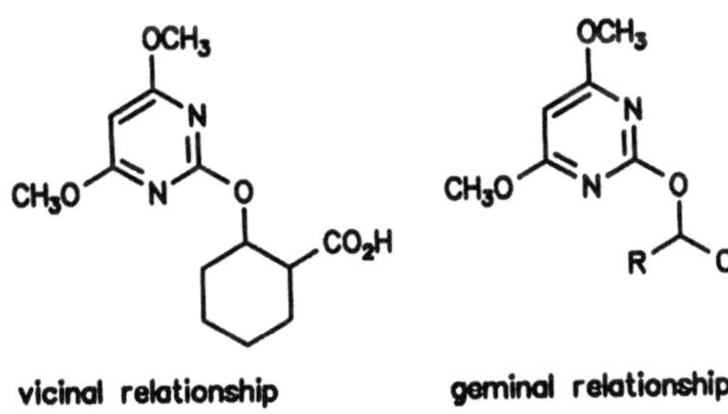

4.2

Various structures have been synthesized and some examples are depicted in Scheme 15, p. 180.
As is the case with compounds described in the previous chapter, it is still too early to postulate any useful structure–activity relationships. Furthermore, it is not known whether these cyclic or acyclic derivatives are better or worse than the original aromatic compounds as regards herbicidal activity.

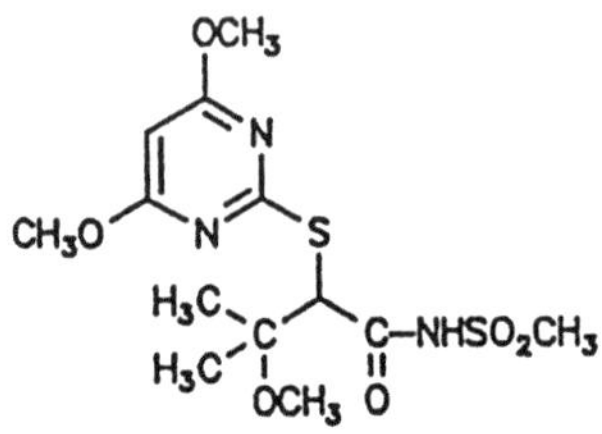

EP 517215 [54]

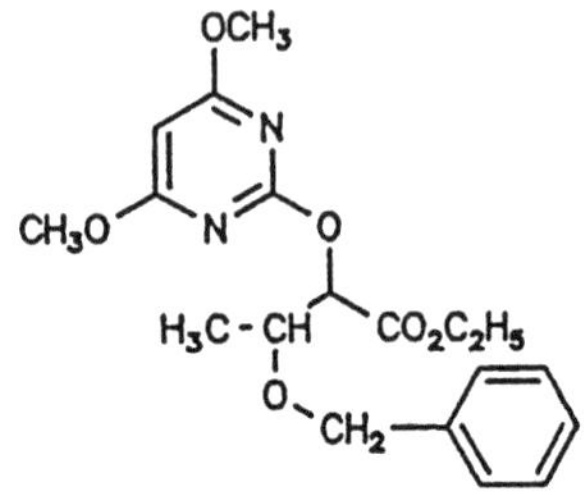

JP 04356470 [55]

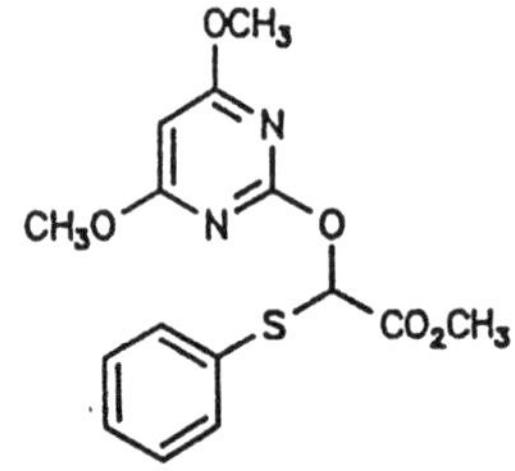

JP 03193765 [58]

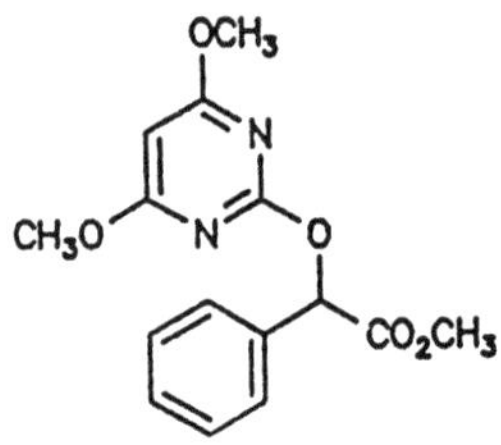

EP 409368 [60]

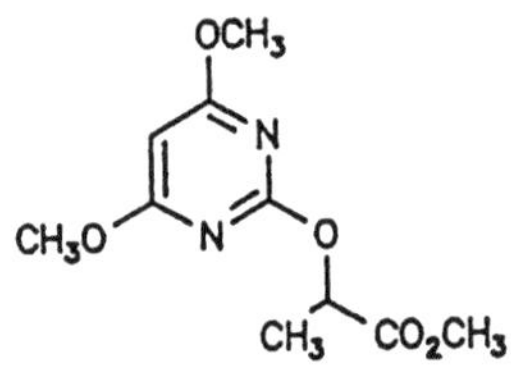

EP 347811 [56]

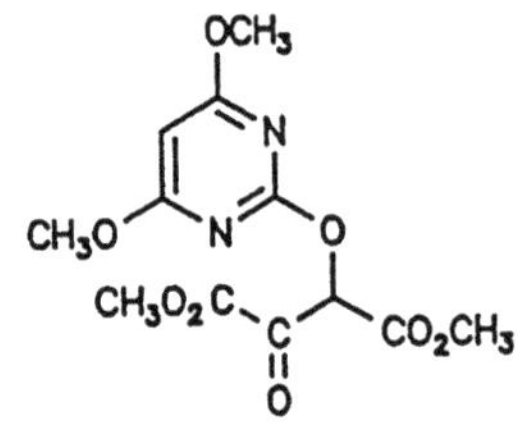

EP 400471 [59]

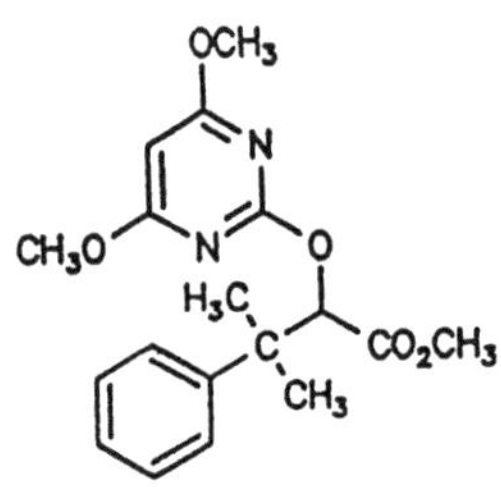

EP 347811 [56]

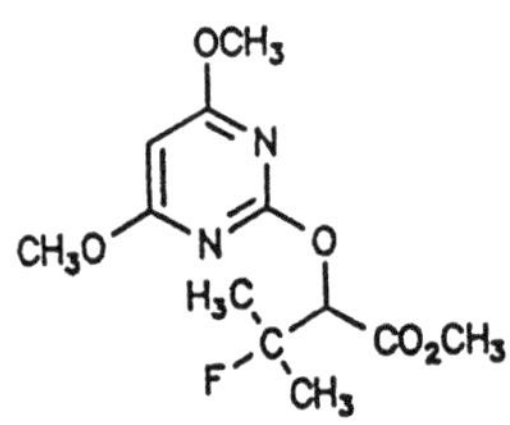

EP 409368 [60]

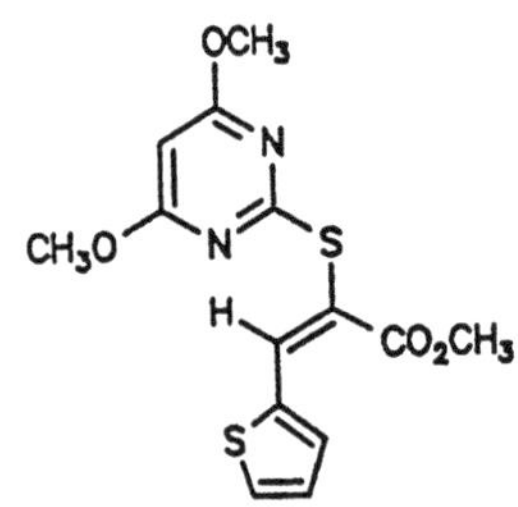

EP 409369 [61]

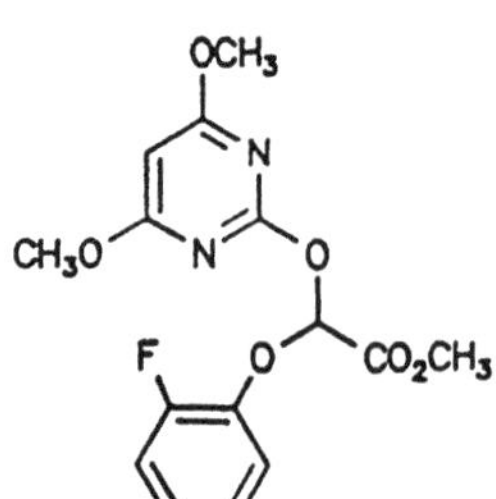

US 5151113 [63]

Scheme 15

5 Miscellaneous Structures

In the previous chapters, compounds of formulae **8** and **9** were described.

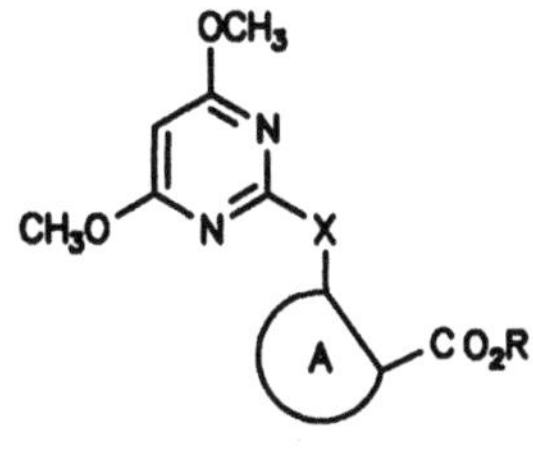

8

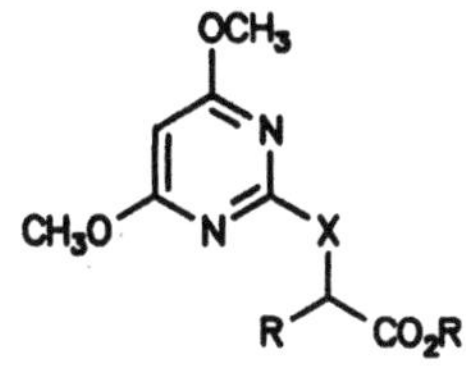

9

In addition to these structures, there are numerous other compounds having a similar structure which mechanistically could also belong to this class of herbicides. Some of these will now be described.

5.1 Pyrimidyl Derivatives

This is a group of compounds in which the acid moiety is bound to a 4,6-dimethoxypyrimidine via a carbon atom and not as in the previous examples by a heteroatom. Some examples [64–66] are illustrated in Scheme 16.

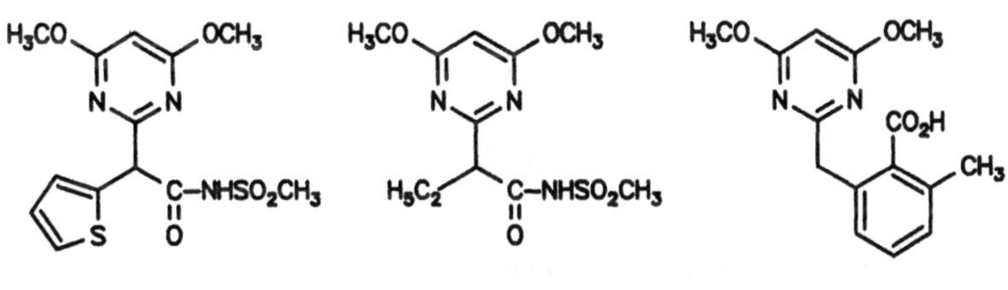

WO 92/01677 [64] JP 04054161 [66] WO 91/10653 [65]

Scheme 16

The synthesis of such compounds generally involves the reaction of an appropriate carbanion with 2-methanesulfonyl-4,6-dimethoxypyrimidine. Such a sequence [65] is shown in Scheme 17, p. 182.

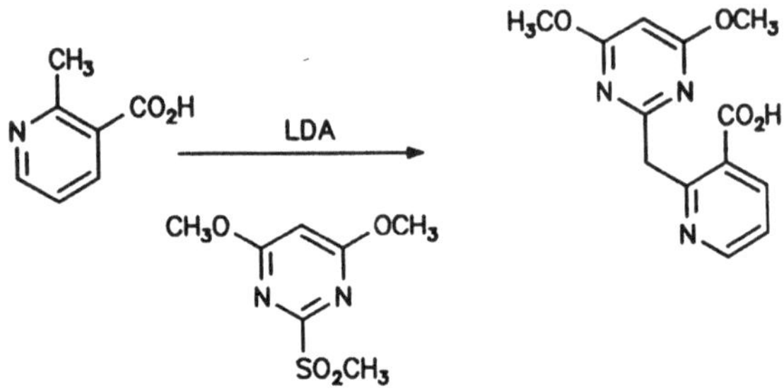

Scheme 17

5.2 Arylpropionic Acid Derivatives

These compounds are very similar to the classical picolinic acid derivatives. The only difference is the insertion of a 2-C unit between the aromatic ring and the acid group. The synthesis of such derivatives [67] involves the reaction of the picolinic acid derivative with 2-methane-sulfonyl-4,6-dimethoxypyrimidine (Scheme 18).

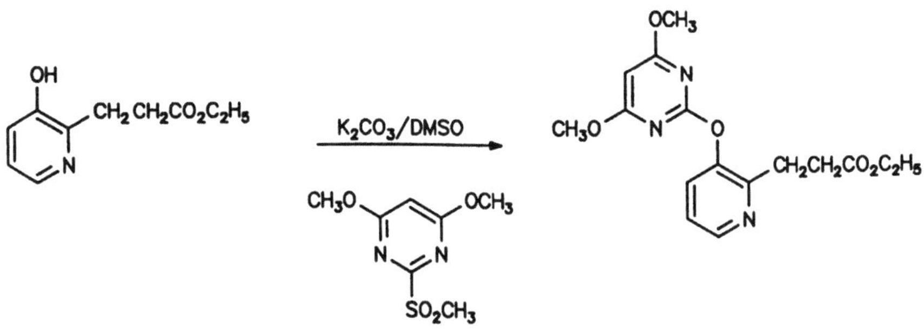

Scheme 18

5.3 Phenoxymethylpyrimidine Derivatives

The insertion of a methylene unit between the oxygen and the 2-position in the pyrimidine ring is a new invention **10** [68].

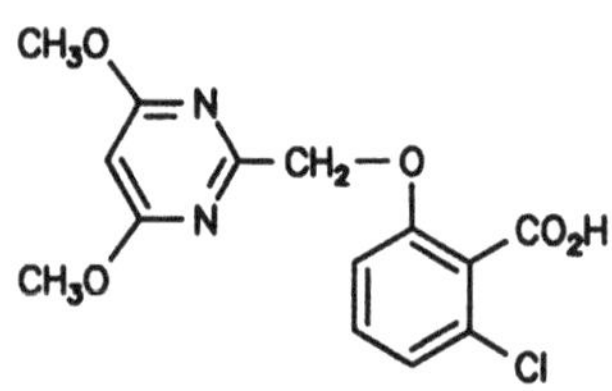

10

Such compounds are probably synthesized by reacting 2-bromomethyl-4,6-dimethyoxypyrimidine with the appropriate salicylic acid derivatives under base catalysis.

5.4 Substituted Aryl and Heteroaryl Phthalide Derivatives

This class of compounds probably also belongs mechanistically to this type of herbicide. Some phthalide derivatives [69] are shown in Scheme 19.

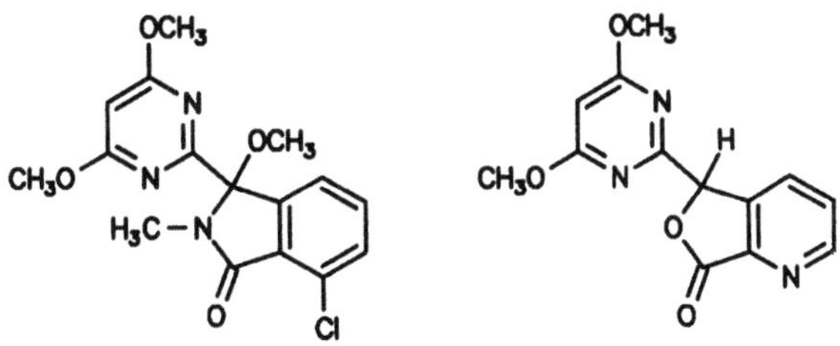

Scheme 19

The synthesis [69] of such substances is depicted in the following sequence (Scheme 20).

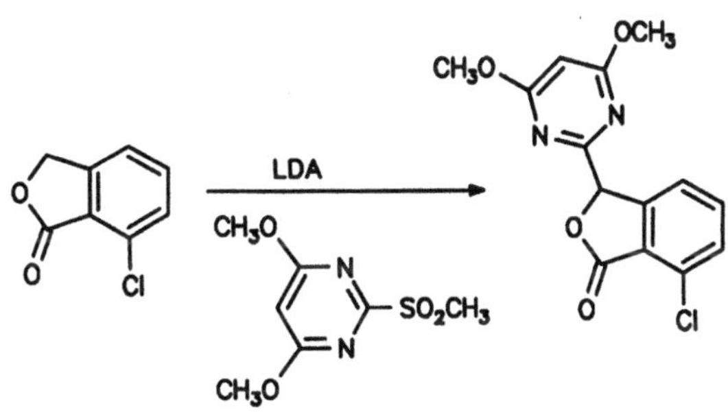

Scheme 20

In all the above cases there is little or no biological data published. It also remains to be seen if these miscellaneous structures really fall into this class of herbicidal compounds.

5.5 A New Tri-Substituted Pyrimidine Derivative

Recently a new pyrimidine [70] of the structure **11** as part of the salicylic acid system has been reported.

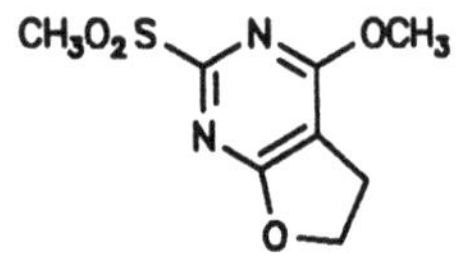

11

It is difficult to judge if this heterocycle has any advantages over the traditional 4,6-dimethoxy-pyrimidine derivative.

6 Mode of Action

At present the only data available [71] is for the compound KIH-2031.

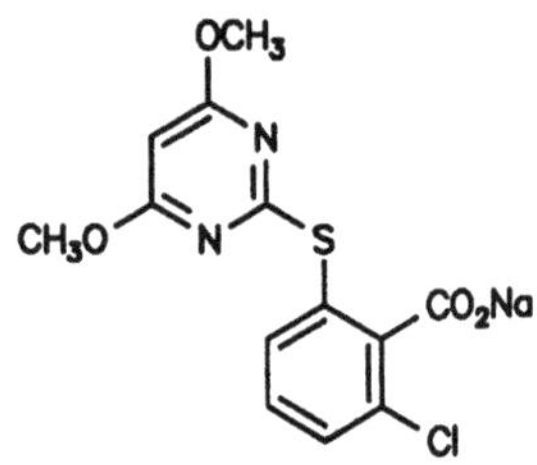

7
KIH–2031

In vitro studies indicate that it inhibits the plant enzyme acetolactate synthase. The level of inhibition is very similar to that of the sulfonylurea Chlorsulfuron (Table 2).

Table 2. Herbicide inhibition of acetolactate synthase (ALS) extracted from rice

Compound	I_{50}^{a} value (mol)
KIH-2031	1.5×10^{-8}
Chlorsulfuron	2.8×10^{-8}

[a] Concentration that causes 50% in vitro inhibition of the acetolactate synthase

7 Experimental Products

Currently, the only known experimental products is KIH-2031. This compound incorporates numerous features which are characteristic for this young class of herbicidal compounds:

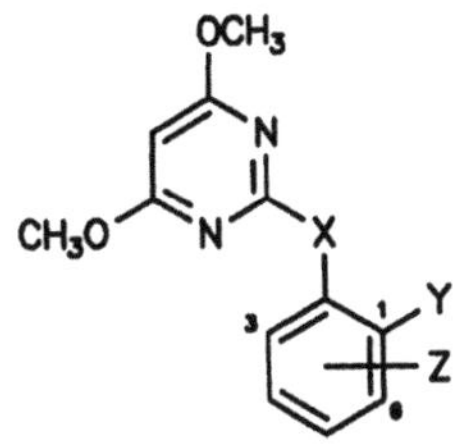

X=S

Y=CO_2Na

Z=6–Cl

12

7.1 Sodium 2-chloro-6-(4,6-dimethoxypyrimidin-2-ylthio) benzoate

KIH-2031 is the first experimental compound originating out of this new class of herbicides.

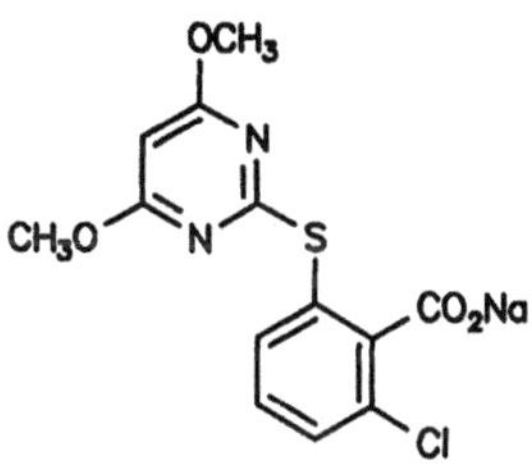

$C_{13}H_{10}ClN_2O_4SNa$
mol. weight 348
m. p. 247.7 °C
LD_{50} acute oral (rat) 1000–3000 mg/kg
LD_{50} acute dermal (rat) > 2000 mg/kg

KIH–2031

13

It is patented [17] by the Kumiai Chemical Industry Company and is being jointly developed under the code KIH-2031/DPX-PE 350 by Kumiai and E.I du Pont de Nemours [71]. KIH-2031 is a highly selective herbicide controlling a wide spectrum of weeds in cotton. It gives excellent control of troublesome weeds such as *lpomoea ssp.*, *Xanthium strumarium*, *Abutilon theophraste*, *Sida spinosa*, *Sesbania exaltata* and *Sorghum halepenses* and can be applied as a pre or post-emergence herbicide. The application rates are in the range 35 to 105 g ai/ha for soil or foliar treatment.

8 References

1. ICI Australia Limited: Europ Pat 1,187 (Australian Prior 13.09.77)
2. Kumiai Chemical Industry Co: Europ Pat 223,406 (Jap Prior 15.10.85)
3. The Annual Meeting of Japanese Pesticide Society: Fukuoka (1992)
4. Sumitomo Chemical Co: Europ Pat 426,476 (Jap Prior 01.11.89)
5. Sumitomo Chemical Co: Europ Pat 457,505 (Jap Prior 15.05.90)
6. Sumitomo Chemical Co: Europ Pat 468,690 (Jap Prior 26.07.90)
7. FMC Corporation US Pat 5149357 (US Prior 19.06.90)
8. Ciba-Geigy AG: Europ Pat 493,321 (British Prior 21.12.90)
9. Sumitomo Chemical Co: Europ Pat 469,711 (Jap Prior 05.07.90)
10. Mitsui Toatsu Chem: Jap Pat 04 221372 (Jap Prior 21.12.90)
11. Mitsui Toatsu Chem: Jap Pat 04 342575 (Jap Prior 17.12.90)
12. Mitsui Toatsu Chem: Jap Pat 03 193766 (Jap Prior 25.12.89)
13. Mitsui Toatsu Chem: Jap Pat 04 342586 (Jap Prior 17.04.91)
14. Mitsui Toatsu Chem: Europ Pat 04 374,839 (Jap Prior 19.12.88)
15. Bayer AG: Europ Pat 459,243 (German Prior 11.04.90)
16. Kumiai Chemical Industry Co: Europ Pat 249,708 (Jap Prior 06.06.86)
17. Kumiai Chemical Industry Co: Europ Pat 315,889 (Jap Prior 04.11.87)
18. BASF: Europ Pat 346,789 (German Prior 16.06.88)
19. Kumiai Chemical Industry Co: Europ Pat 287,072 (Jap Prior 14.04.87)
20. Kumiai Chemical Industry Co: Europ Pat 287,079 (Jap Prior 14.04.87)
21. Kumiai Chemical Industry Co: Europ Pat 321,846 (Jap Prior 22.12.87)
22. Kumiai Chemical Industry Co: Jap Pat 05 32639 (Jap Prior 31.07.91)
23. Kumiai Chemical Industry Co: Europ Pat 472,113 (Jap Prior 22.08.90)
24. Kumiai Chemical Industry Co: Jap Pat 04 112876 (Jap Prior 30.08.90)
25. Kumiai Chemical Industry Co: Europ Pat 435,170 (Jap Prior 28.12.89)
26. Kumiai Chemical Industry Co: Jap Pat 05 32638 (Jap Prior 31.07.91)
27. BASF: Europ Pat 402,751 (German Prior 14.06.89)
28. BASF: Europ Pat 414,058 (German Prior 19.08.89)
29. BASF: Europ Pat 435,104 (German Prior 22.12.89)
30. BASF: Europ Pat 490,060 (German Prior 26.10.90)
31. BASF: Europ Pat 527,380 (German Prior 10.08.91)
32. BASF: Europ Pat 527,378 (German Prior 10.08.91)
33. BASF: German Pat 4,030,929 (German Prior 29.09.90)
34. Bayer AG: Europ Pat 477,637 (German Prior 22.09.90)
35. Kumiai Chemical Industry Co: Europ Pat 249,707 (Jap Prior 14.06.86)
36. Kumiai Chemical Industry Co: Europ Pat 330,990 (Jap Prior 22.02.88)
37. Kumiai Chemical Industry Co: Europ Pat 335,409 (Jap Prior 31.03.88)
38. Kumiai Chemical Industry Co: Europ Pat 507,962 (Jap Prior 25.10.90)
39. BASF: Europ Pat 360,163 (German Prior 22.09.88)
40. Bayer AG: Europ Pat 472,925 (German Prior 10.08.90)
41. Mitsui Toatsu Chem: Europ Pat 435,186 (Jap Prior 26.12.89)
42. Kumiai Chemical Industry Co: Jap Pat 01 290671 (Jap Prior 16.05.88)
43. Kumiai Chemical Industry Co: Jap Pat 04 145081 (Jap Prior 10.06.90)
44. Kumiai Chemical Industry Co: Jap Pat 03 284676 (Jap Prior 30.03.90)
45. Kumiai Chemical Industry Co: Jap Pat 03 232884 (Jap Prior 07.02.90)
46. Kumiai Chemical Industry Co: Jap Pat 02 056469 (Jap Prior 08.12.87)
47. Mitsubishi Petrochemical Co: Europ Pat 468,766 (Jap Prior 24.07.90)
48. Nissan Chemical Industry: US Pat 4973354 (Jap Prior 01.12.88)
49. Shell Int Research: Europ Pat 431,707 (British Prior 08.12.89)
50. Schering AG: Europ Pat 439,243 (German Prior 31.05.91)

51. Kumiai Chemical Industry Co: Jap Pat 04 356469 (Jap. Prior 31.05.91)
52. Shell Int Research: WO 92/22538 (Prior 10.06.91)
53. Ube Industries: Europ Pat 481,512 (Jap Prior 19.10.90)
54. Ube Industries: Europ Pat 517,215 (Jap Prior 07.06.91)
55. Ube Industries: Jap Pat 04 356470 (Jap Prior 31.05.91)
56. Kumiai Chemical Industry Co: Europ Pat 347,811 (Jap Prior 20.06.88)
57. Kumiai Chemical Industry Co: Jap Pat 01 301668 (Jap Prior 30.05.88)
58. Kumiai Chemical Industry Co: Jap Pat 03 193765 (Jap Prior 22.12.89)
59. Shell Int Research: Europ Pat 400,471 (British Prior 02.06.89)
60. Schering AG: Europ Pat 409,368 (German Prior 19.07.89)
61. Schering AG: Europ Pat 409,369 (German Prior 19.07.89)
62. Schering AG: Europ Pat 422,751 (German Prior 09.10.89)
63. Dow Elanco: US Pat 5151113 (US Prior 29.04.91)
64. Schering AG: WO 92/01677 (Prior 12.07.91)
65. EI Du Pont De Nemours: WO 91/10653 (Prior 11.01.90)
66. Kumiai Chemical Industry Co: Jap Pat 04 054161 (Jap Prior 25.06.90)
67. Hokko Chemical Industry: Jap Pat 04 059980 (Jap Prior 25.05.90)
68. Mitsubishi Petrochemical Co: WO 92/19603 (Prior 30.04.91)
69. Sandoz: Europ Pat 461,079 (US Prior 07.06.90)
70. BASF: Europ Pat 490,224 (German Prior 14.12.90)
71. Nezu M, Claus JS, Williams CS (1991) Brighton Crop Protection Conference-Weeds p 57

Miscellaneous Structures

F. Lieb and U. C. Philipp

Bayer AG, ZF-FWI, D-51368 Leverkusen, FRG

The modification of the sulfonylurea entity has led to promising new substances. The rigidification of the urea bridge or its replacement with heterosubstituted guanidines and isothioureas has brought about new herbicidal compounds. Acylated (het)aryl sulfonamides – either with conventional pyrimidine or triazine moieties or new heterocyclic systems – and sulfamoylsulfonylureas are also herbicidally active.

Chemistry of Plant Protection, Vol. 10

Miscellaneous Structures

For many years it seemed that the herbicidal activity of the sulfonylureas was severely reduced or even lost if one of the three structural elements – (het)aryl part, sulfonylurea bridge or pyrimidine/triazine moiety – was omitted or replaced with a different functionality.

In recent years, however, many patents have been published which claim herbicidal activity for considerably modified yet still sulfonylurea-related structures.

This contribution compiles these putatively ALS-inhibiting "miscellaneous structures" and tries to show their relationships to the classical sulfonylureas. The first two sections report on modifications of the urea bridge. The third paragraph deals with alternatives to the pyrimidine/triazine ring. The final part lists replacements for the (het)aryl moiety.

1 Bridged Sulfonylureas

Quite frequently, rigidification of structures leads to enhanced biological activity of substances. This hypothesis suggests the incorporation of the urea bridge into a ring system. Accordingly, one can envisage N–C–X (× × × ×), N′–C–X (○○○○) or N–C–N′ (+ + + +) bridging, respectively, which is shown in **1**. This section provides examples for each type of urea bridging.

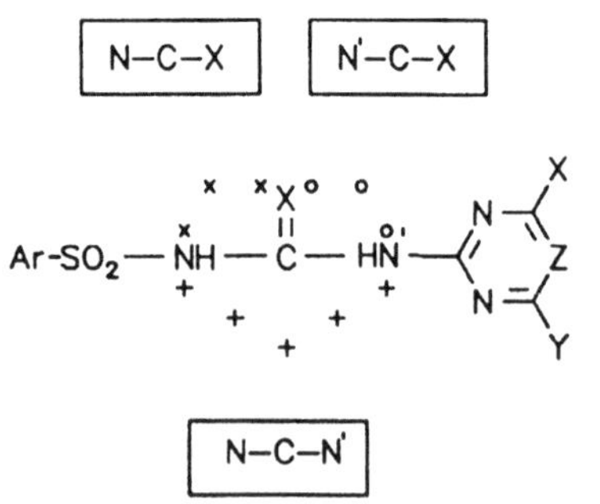

1

1.1 N′–C–X Bridging

Bond formation between N′ and X in **1** (○○○○) requires the transition to iso(thio)ureas e.g. thiazolinone (**3**), which can be obtained by condensation of the classical sulfonylthiourea (**2a**) and chloroacetyl chloride (Scheme 1) [1]. Similar compounds such as **4** [2] and **5** [3] have also been claimed as herbicides.

The incorporation of the urea into an aromatic ring is exemplified by sulfonyliminothiadiazoline (**9**). Scheme 2 displays its concise two-step synthesis by coupling sulfonyl chloride (**6**), thiadiazole (**7**) and pyrimidine (**8**) [4].

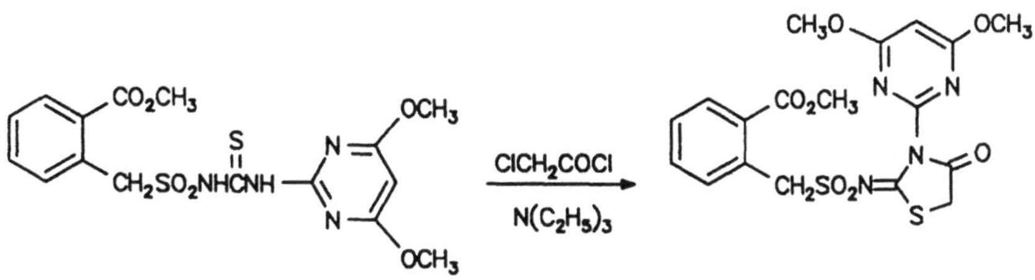

2a **3**

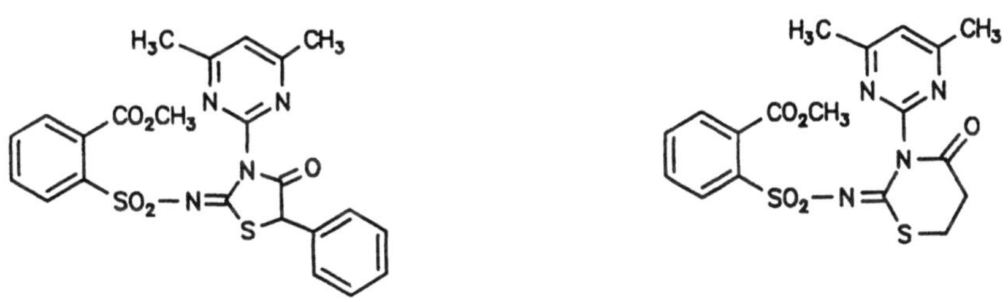

4 **5**

Scheme 1

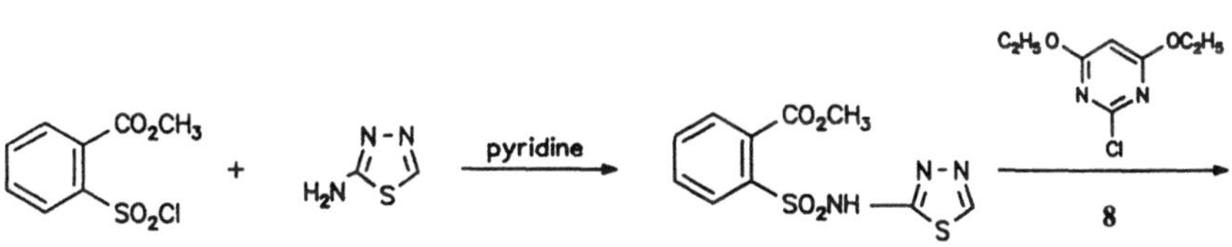

6 **7**

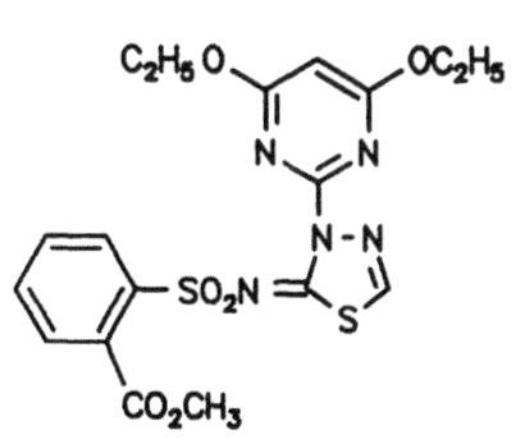

9

Scheme 2

Arrangement of sulfamoyl- and pyrimidine residue in 1,3- rather than 1,2-fashion furnishes triazoles (**10**) (N-linked) [5], **(11)** (C-linked) [6] or **(12)** (N-linked, substituted) [7], which are obtained according to Schemes 3–5, respectively.

Inversion of the SO_2NH-link between benzene and triazole in **13** leads to sulfanilide (**14**) which structurally resembles the triazolopyrimidines (see contrib.

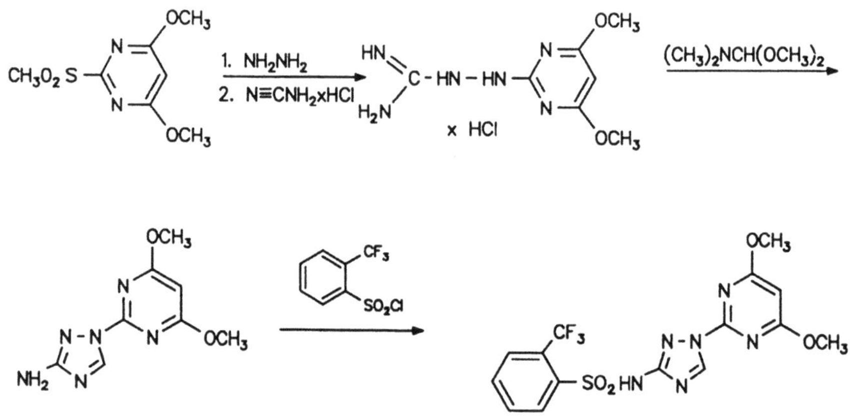

10

Scheme 3

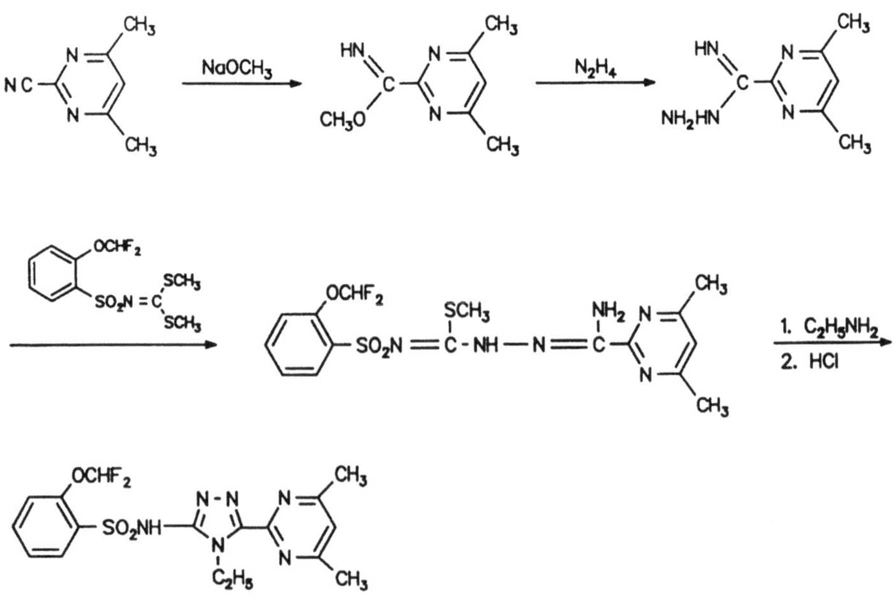

11

Scheme 4

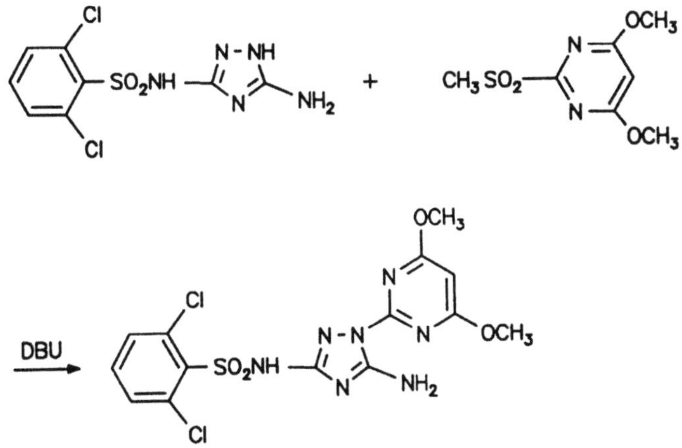

12

Scheme 5

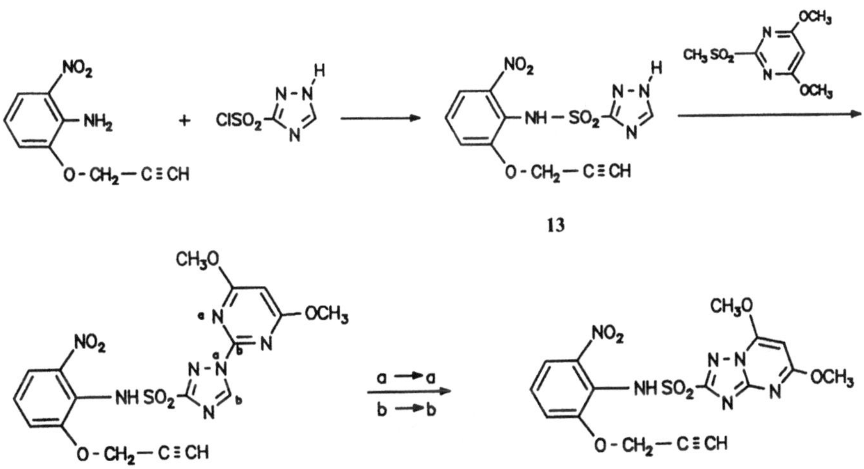

13 14 15

Scheme 6

of W. A. Kleschick, this volume): matching the atoms marked a and b translates pyrimidyl triazole (**14**) into triazolopyrimidine (**15**) (Scheme 6) [8].

1.2 N–C–X Bridging

Again, incorporation of N and X in **1** (× × × ×) into a ring system leads to iso(thio)ureas. 1,3-Thioacetidine (**17**) is obtained when the classical sulfonylurea (**16**) is cyclized with diiodomethane according to Scheme 7 [9].

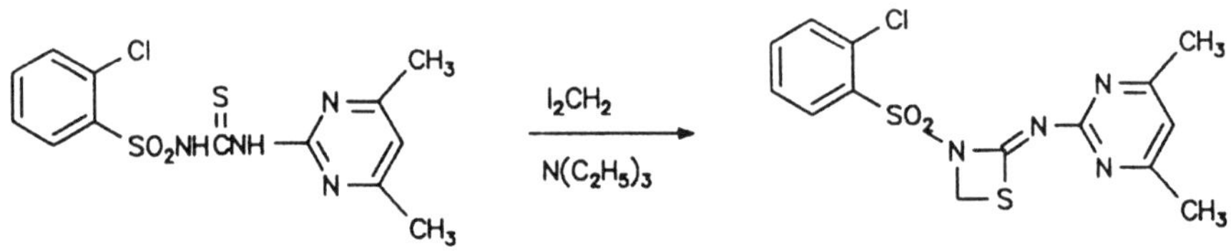

16 **17**

Scheme 7

This kind of bridging has also been realized for thiadiazolines, e.g. **19**, which is assembled from thiadiazole (**7**), pyrimidine (**18**) and sulfonyl chloride (**6**) (Scheme 8) [10], similarly to **9** (Scheme 2).

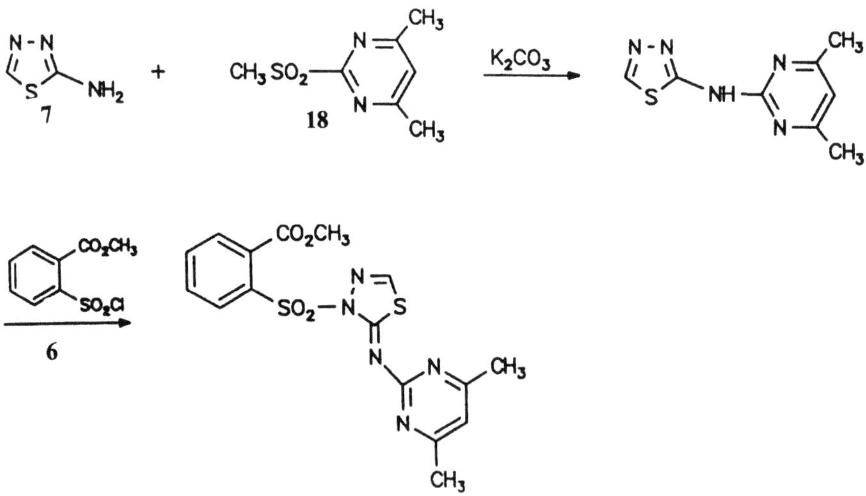

19

Scheme 8

1.3 N–C–N′ Bridging

Urea cyclization via the nitrogen atoms of **1** (+ + + +) has been accomplished in hexahydrotriazinones and triazolinones [11, 12]. Bensulfuron methyl (**2b**) can be rigidified by means of a Mannich reaction to give hexahydrotriazinone (**20**) according to Scheme 9.

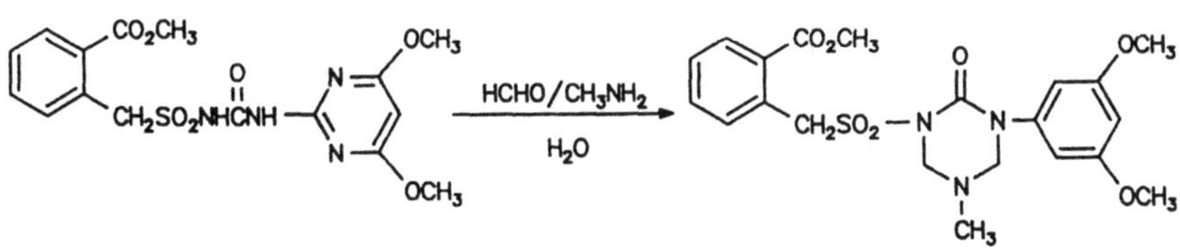

2b 20

Scheme 9

2 Heterosubstituted Sulfonyl Guanidines and Isothioureas

As shown in the contribution of J. V. Hay et al., this volume, the sulfonylurea bridge may be considerably modified without dramatic impact on the degree of ALS-inhibition. Thus, thioureas, guanidines and the respective isostructures as well as "spacered" ureas (e.g. bensulfuron methyl) rank amongst the potent ALS-inhibitors. This section focusses on heterosubstituted guanidines and isothioureas which have recently been studied in great detail.

2.1 Sulfonylaminoguanidines

According to Scheme 10, sulfonyl isothiourea (**21b**) [13, 14] reacts smoothly with hydrazine to yield aminoguanidine (**22a**) [15]. This high susceptibility towards hydrazinolysis could mean that sulfonylisothioureas represent prodrugs of classical sulfonylureas which would be formed by hydrolysis of, e.g. **21a**.

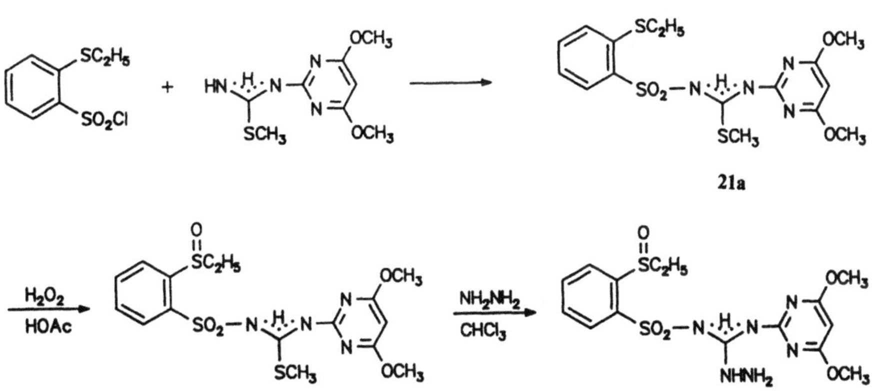

21b 22a

Scheme 10

Alternatively, pyrimidyl cyanamide (**23**) adds methoxyl amine to form **24** and then condenses with two equivalents of sulfonyl chloride (**25**) to give bissulfonylmethoxyguanidine (**26a**) – again potentially a prodrug – which can be readily hydrazinolyzed to **22b** (Scheme 11) [16].

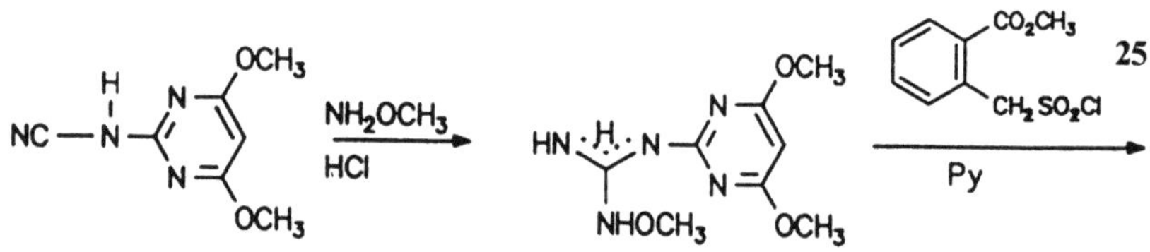

23 **24a**

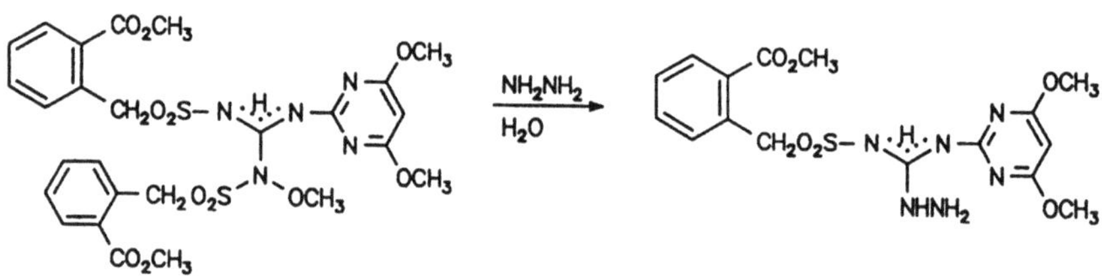

26a **22b**

Scheme 11

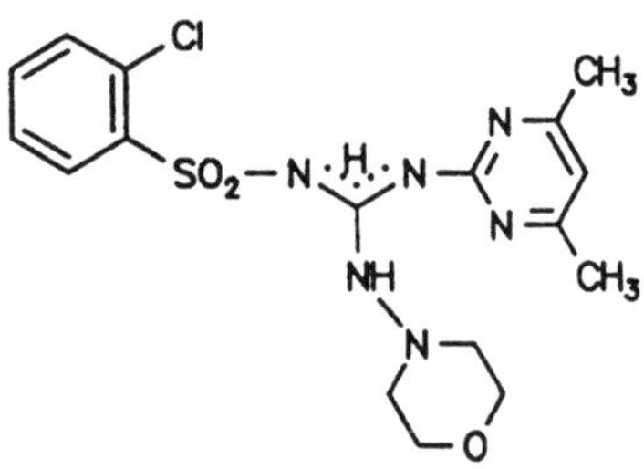

22c

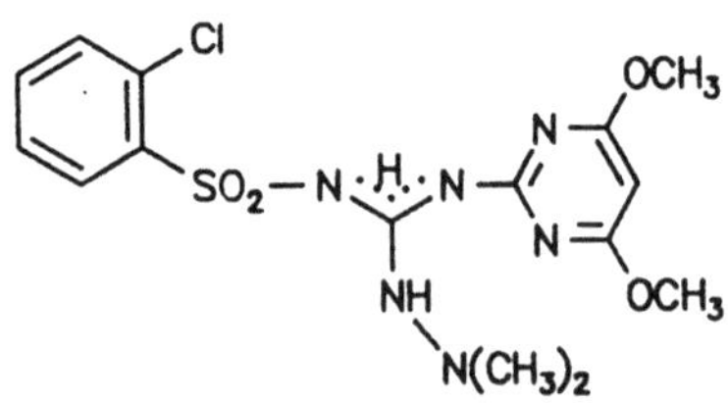

22d

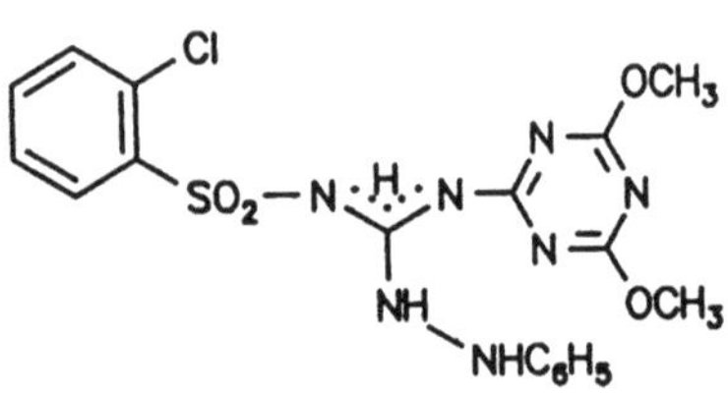

22e

Scheme 12

Most substituted hydrazines react analogously as indicated by the representative examples **22c** [17], **22d** [18] and **22e** [19] in Scheme 12.

Due to the excellent leaving group properties of the sulfonyl-*O*-methylhydroxylamino moiety even pyrazole (**27**) attacks **26b** to form pyrazolyl amidine (**28**) [20] (Scheme 13). Similar imidazolyl and triazolyl amidines have also been claimed [21].

Sulfonylaminoguanidines may be transformed into the respective hydrazones (**29**) by reaction with appropriate aldehydes (**29a**) or ketones (**29b**) [15]. Again, these derivatives, e.g. **29c**, can also be obtained from the versatile precursor **26b** [22] (Scheme 14).

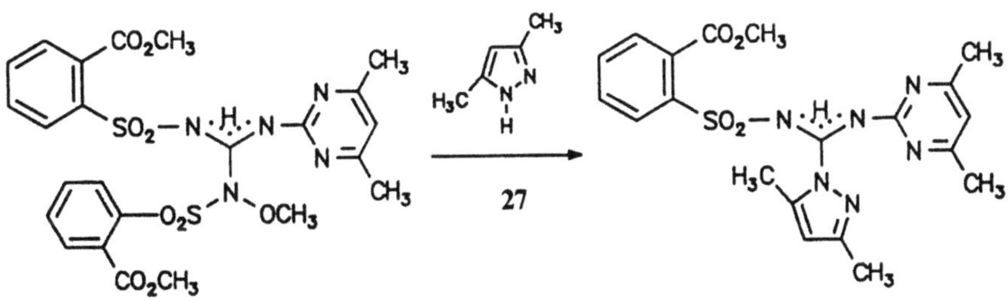

26b **28**

Scheme 13

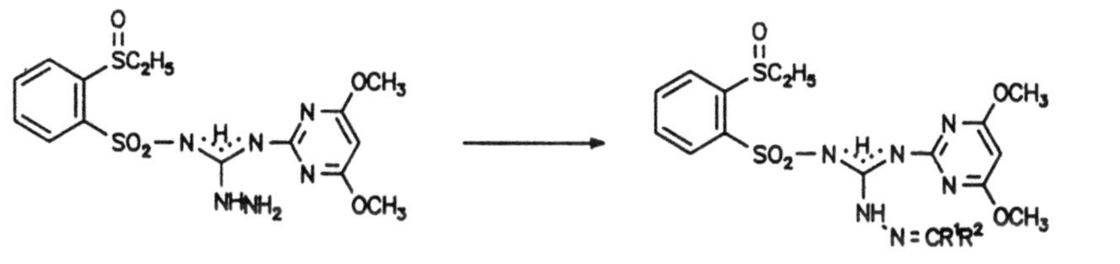

22a

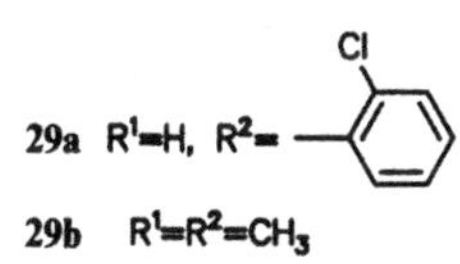

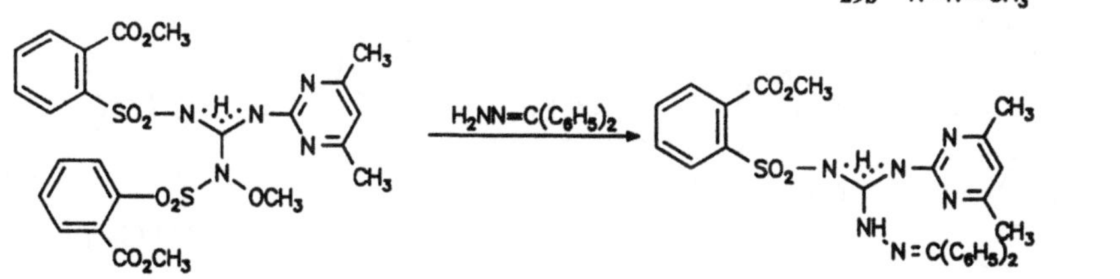

26b **29c**

Scheme 14

Condensation of hydrazine (**22f**) with DMF-acetal gives access to amidine (**30**) [22] (Scheme 15).

Sulfonation of hydrazine (**22b**) furnishes sulfonyl hydrazide (**31**) [23], which is a more potent herbicide than comparable acylated aminoguanidines like **32**, once more accessible via relay compound **26c** [17, 24] (Scheme 16).

Finally, isocyanates can be added to the guanidine NH-function of dialkylaminoguanidines like **22g** to furnish urea (**33**) [25] (Scheme 17).

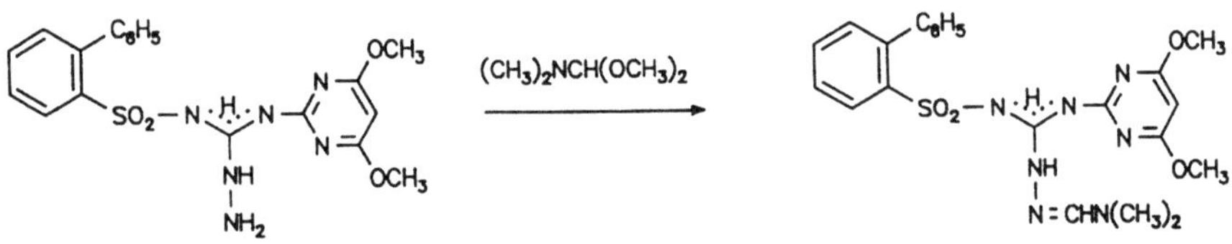

22f **30**

Scheme 15

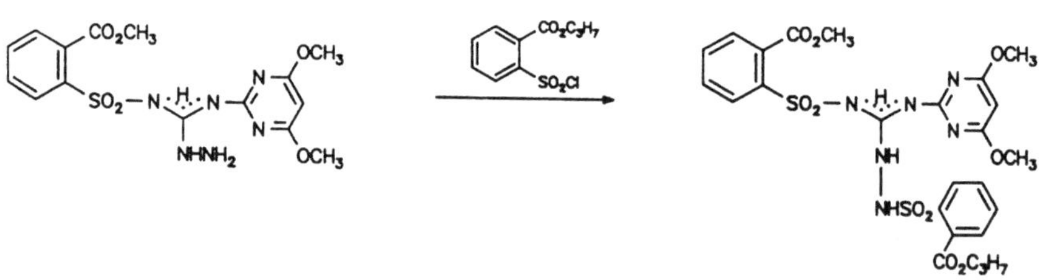

22b **31**

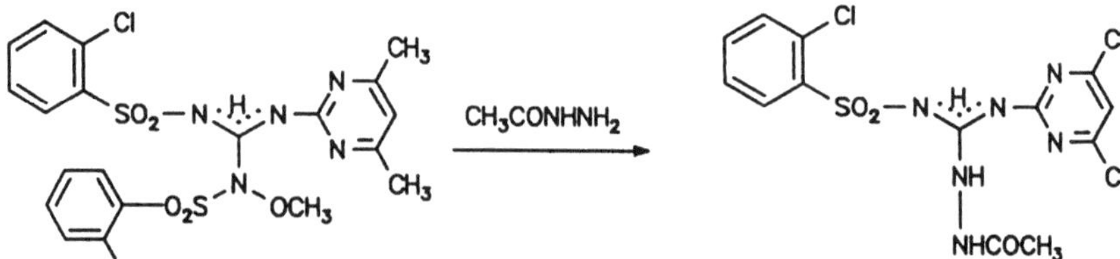

26c **32**

Scheme 16

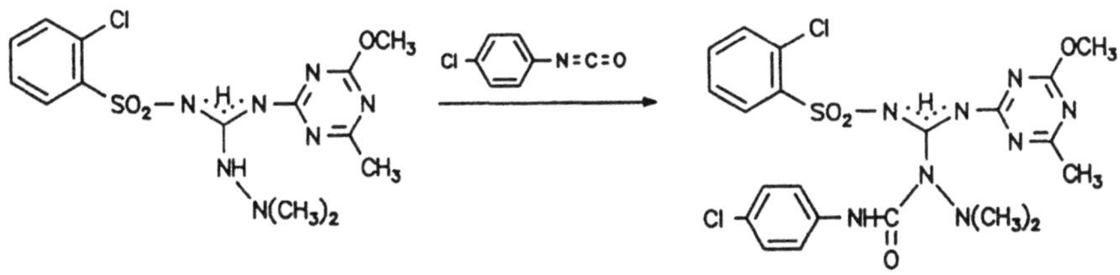

22g

33

Scheme 17

2.2 Sulfonylhydroxyguanidines

Replacement of the keto group with an oxime function was among the first structural modifications of sulfonylureas [26, 27]. As for sulfonylaminoguanidines (**22**), two synthetic approaches to sulfonylhydroxyguanidines (**34**) and (**35**) are important, starting from *S*-methylisothiourea (**21**) [26, 28] or bissulfonylmethoxyguanidine (**26c**) [17], respectively, (Scheme 18).

Cyclic alkoxyamines such as isoxazolidine (**36**) may be used analogously to obtain isoxazolidinyl amidine (**37**), which is also formed by addition of isoxazolidine to triazinyl cyanamide (**38**) and consecutive sulfonylation [29] (Scheme 19).

The sulfonylhydroxy- or -alkoxyguanidines (**34**) and (**35**) can be broadly derivatized with the herbicidal activity being preserved. Acylation of (**34**) with chloroformate furnishes carbamate (**39**) (Scheme 20). Similar reactions with acid chlorides and isocyanates have been claimed [27].

Bissulfonylalkoxyguanidines of type **26** have already been mentioned as intermediates for various sulfonyl guanidine derivatives [30, 31]. Due to their fairly high reactivity one might argue that these compounds act as prodrugs of classical sulfonylureas. Different sulfonyl residues may be introduced through sulfonation of sulfonylalkoxyguanidine (**35c**) to yield **26d** (Scheme 21).

Alternatively, pyrimidinylalkoxyguanidine (**40**) can be sulfonated stepwise to form **26e** via **41** [32] (Scheme 22).

Similarly, acylation of sulfonylmethoxyguanidine (**35d**) by 4-chlorobenzoyl chloride results in benzoylmethoxyguanidine (**42**) [25] (Scheme 23).

2.3 Bissulfonylguanidines

The synthesis of bissulfonylguanidines, e.g. **44**, is easily accomplished by addition of aniline to pyrimidinyl cyanamide (**43**) and consecutive sulfonylation [33] (Scheme 24).

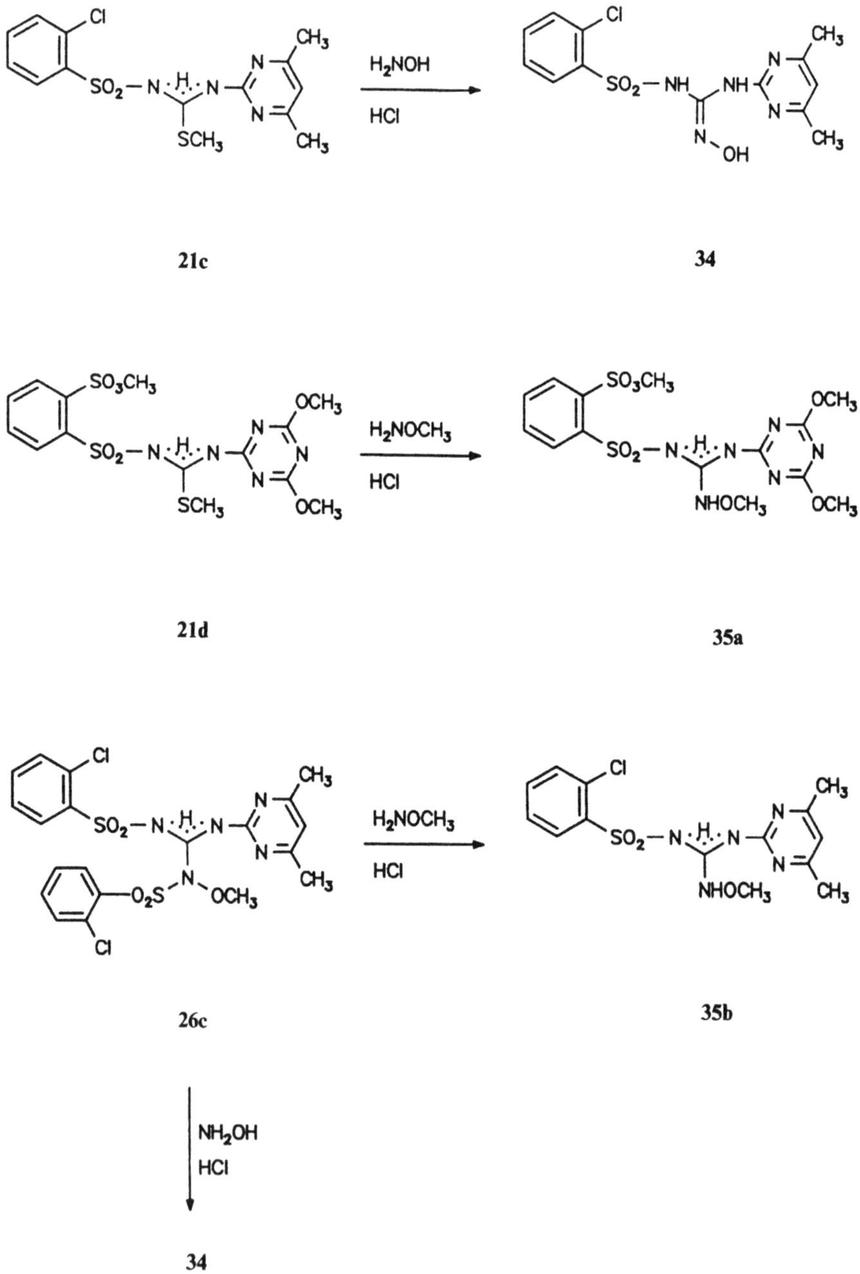
H2NOH
HCl
21c
34
H2NOCH3
HCl
21d
35a
H2NOCH3
HCl
26c
35b
NH2OH
HCl
34

Scheme 18

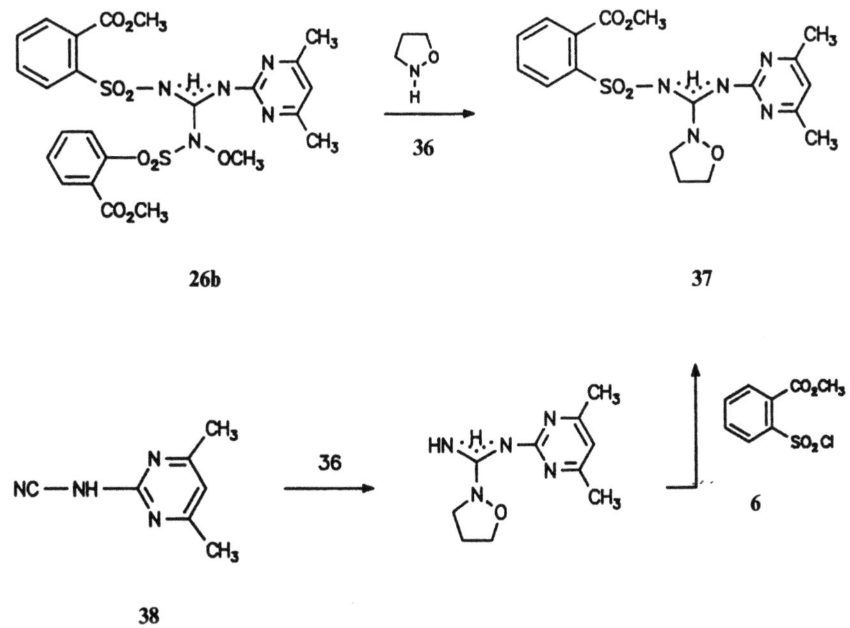

Scheme 19

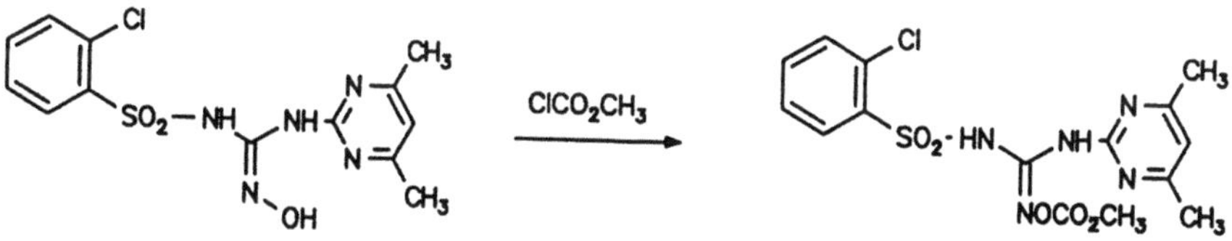

34

39

Scheme 20

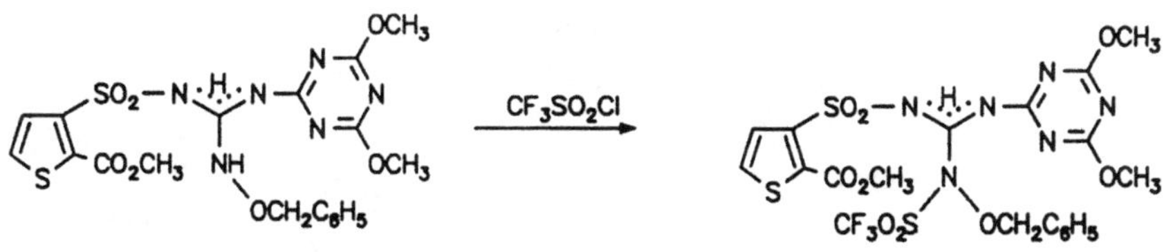

35c

26d

Scheme 21

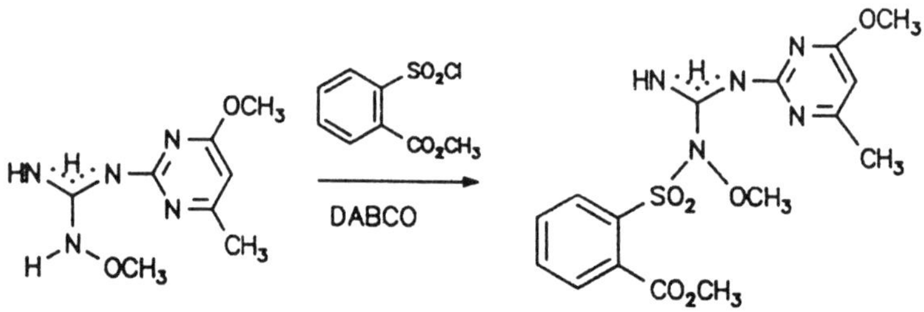

40

41

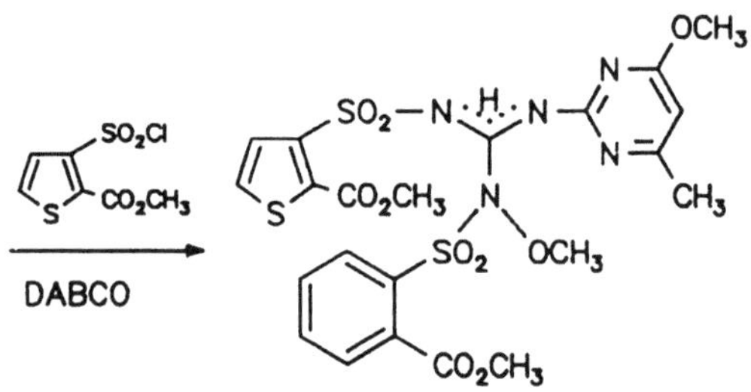

26e

Scheme 22

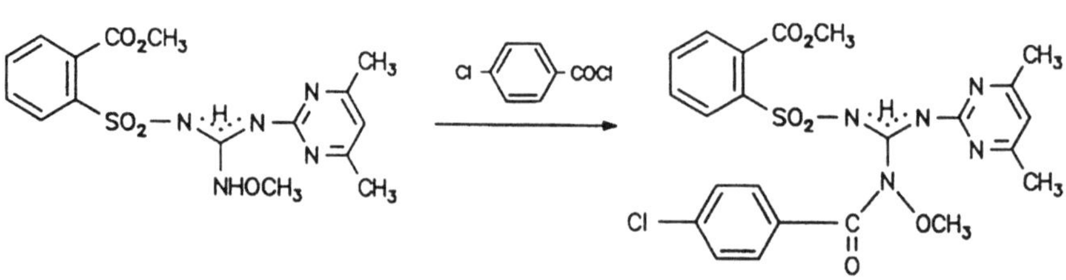

35d

42

Scheme 23

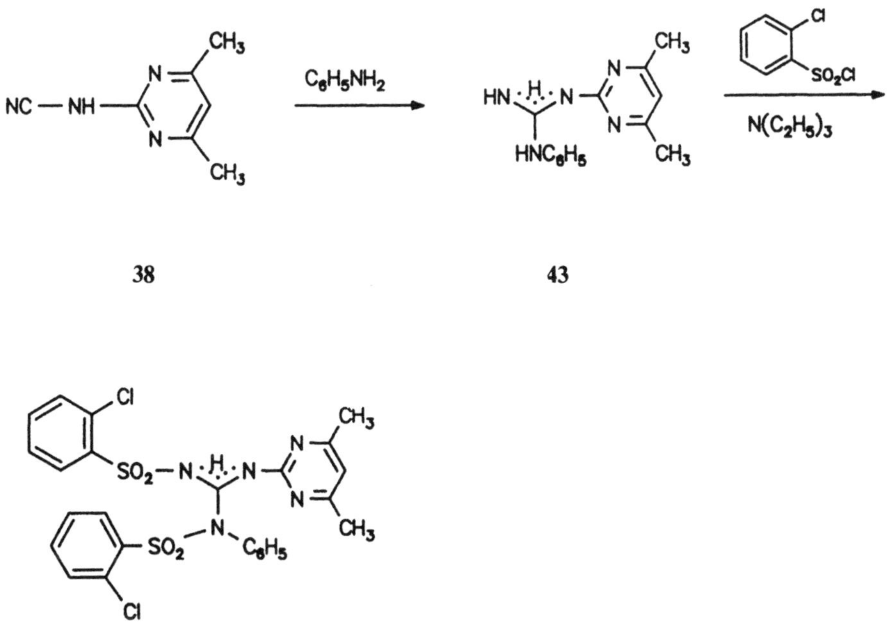

38 43

44

Scheme 24

2.4 Sulfonylisothioureas

Sulfonylisothiourea disulfides such as **47** have been claimed to be herbicidally active. Symmetric disulfides are prepared by oxidative dimerisation [35], whereas unsymmetric disulfides are synthesized through oxidative coupling if sulfonyl thiourea (**45a**) and *S*-alkyl thiourea (**46**) [34] (Scheme 25) are used.

3 Acylated Sulfonamides

This section deals with the modification of the aminopyrimidine/-triazine moiety of classical sulfonyl ureas. In sect. 3.1, "true" acylated sulfonamides **(48)** (C-linked), are reported which formally originate from the omission of the N′H-function of classical sulfonyl ureas. Section 3.2 is concerned with N-linked acylated sulfonamides **(49)** which could also be classified as *N′*, *N′*-bridged sulfonyl ureas (see sect. 1).

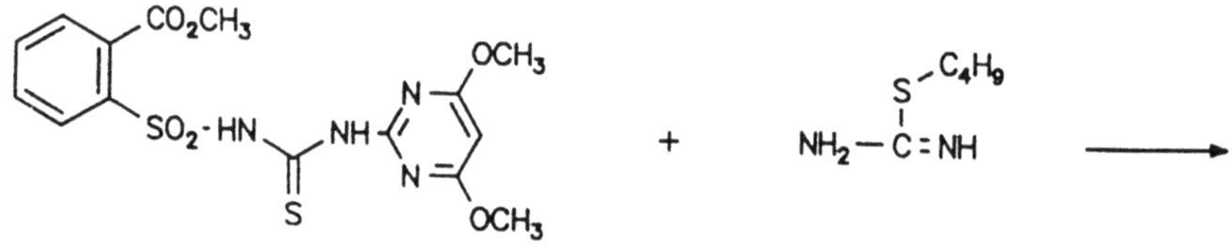

45a 46

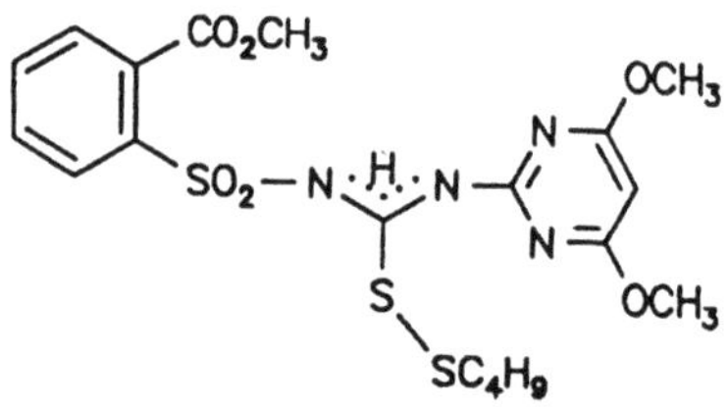

47a

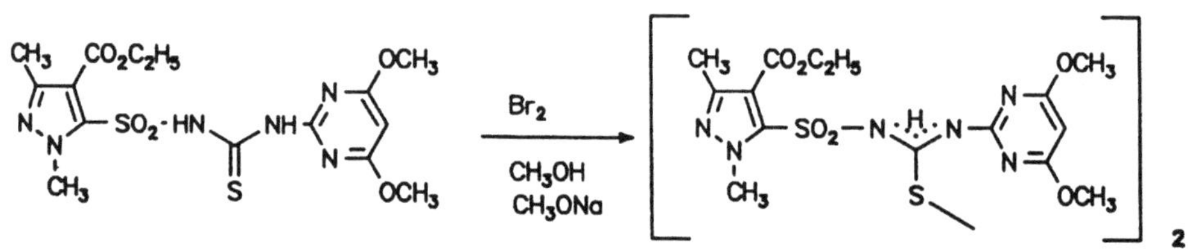

45b 47b

Scheme 25

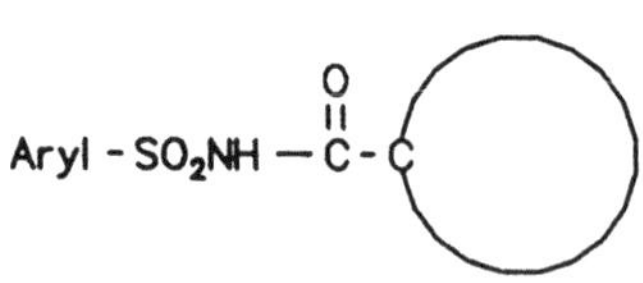

48

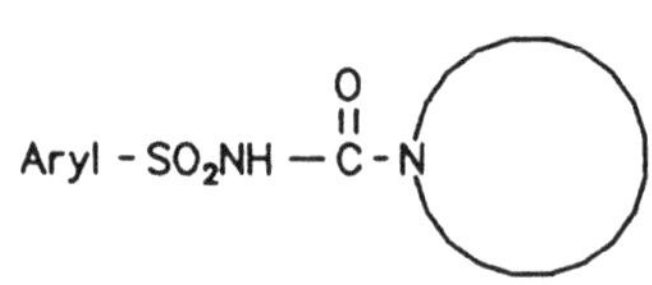

49

3.1 Hetaryl–C–carbonyl Sulfonamides

The transition from sulfonylureas to sulfonamides requires the synthesis of adequate heterocyclic carboxylic acids as shown in Scheme 26. Pinner reaction of malononitrile (**50**), cyclization of the resulting iminoester salt (**51**) with ethyl

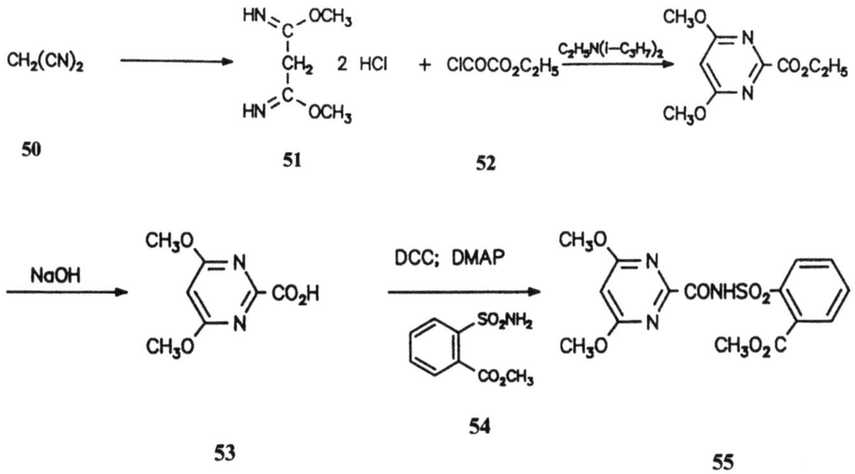

Scheme 26

oxalyl chloride (**52**) and saponification leads to pyrimidine carboxylic acid (**53**). Coupling of **53** to sulfonamide (**54**) furnishes **55** which is just one NH-group shorter than sulfometuron, when minor discrepancies in the pyrimidine substitution are ignored.

The extension of **55** by one methylene unit (cf. sulfometuron → bensulfuron-methyl) gives **56** which can be looked upon as the carbon analogue of sulfometuron [36] (Scheme 27). The anilide (**57**) represents sulfometuron with the sulfonyl bridge being permuted, the phenolic sulfamate (**58**) is just the oxo-analogue thereof. Extending the alienation slightly farther, one gets to anilide (**59**) and its carbamate (**60**) by interchange of two pyrimidine substituents [37].

Even 5-substituted pyrimidines such as **61** [38] and sulfonamides acylated by imidazole- or pyrazole carboxylic acids (**62**) or (**63**) [39], respectively, are still herbicidally active. Scheme 28 displays two options for the synthesis of **62** and **63**.

The idea of rigidification and isosteric modification has probably led to bicyclic acylated sulfonamides (**64**–**66**). While the distance between the arene and the pyrimidine moieties remains unchanged, the urea N′H-group has been replaced with an sp^2-carbon of the respective azole ring which additionally fixes the planarity in this part of the molecule. The structural vicinity to the triazolopyrimidines of W. A. Kleschick's contribution to this volume is also obvious. A short synthesis of **64** is given in Scheme 29 [40].

Sulfonylcarboxamides related to imidazolinones are discussed in Sect. 4.2.2.

Finally, acylated methane sulfonamides (**67**) and (**68**), which show only remote structural relationship with classical sulfonylureas, have attracted attention as herbicides. They can be prepared as outlined in Scheme 30 [41].

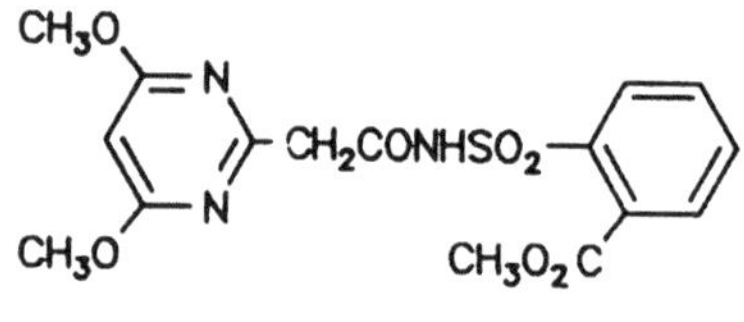

56

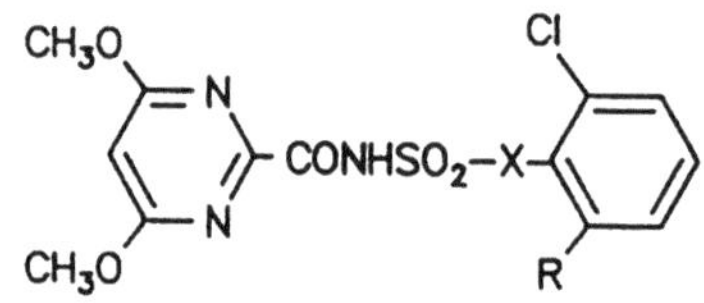

57 X=NH, R=H

58 X=O, R=Cl

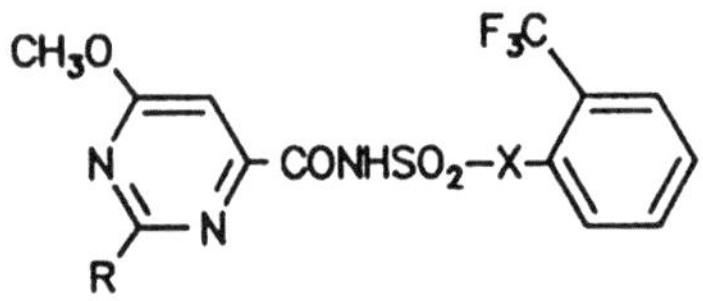

59 X=NH, R=OCH_3

60 X=CH_2, R=CH_3

Scheme 27

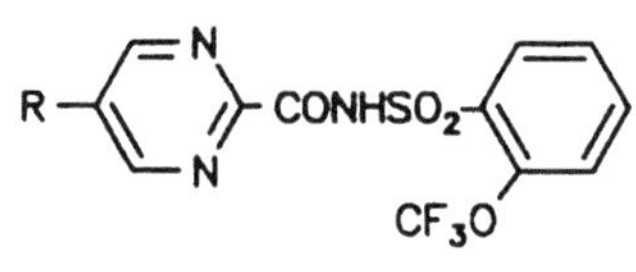

61a: R=OCH_3

61b: R=CH_3

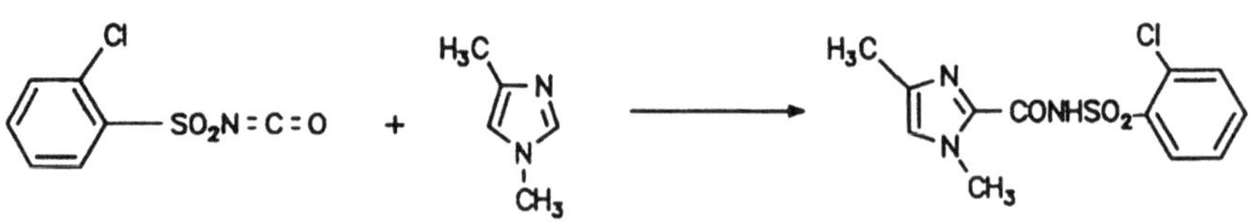

62

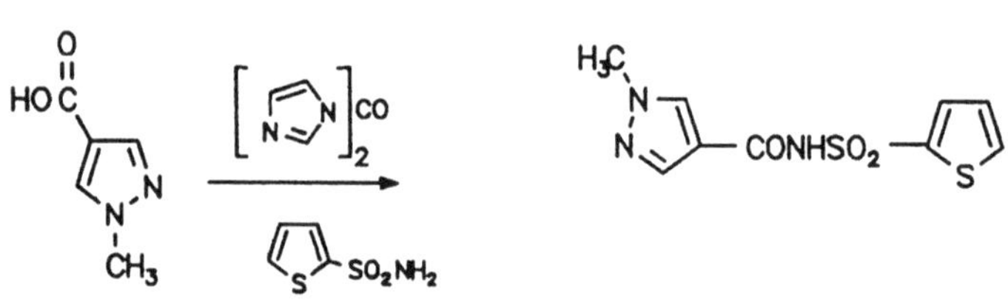

63

Scheme 28

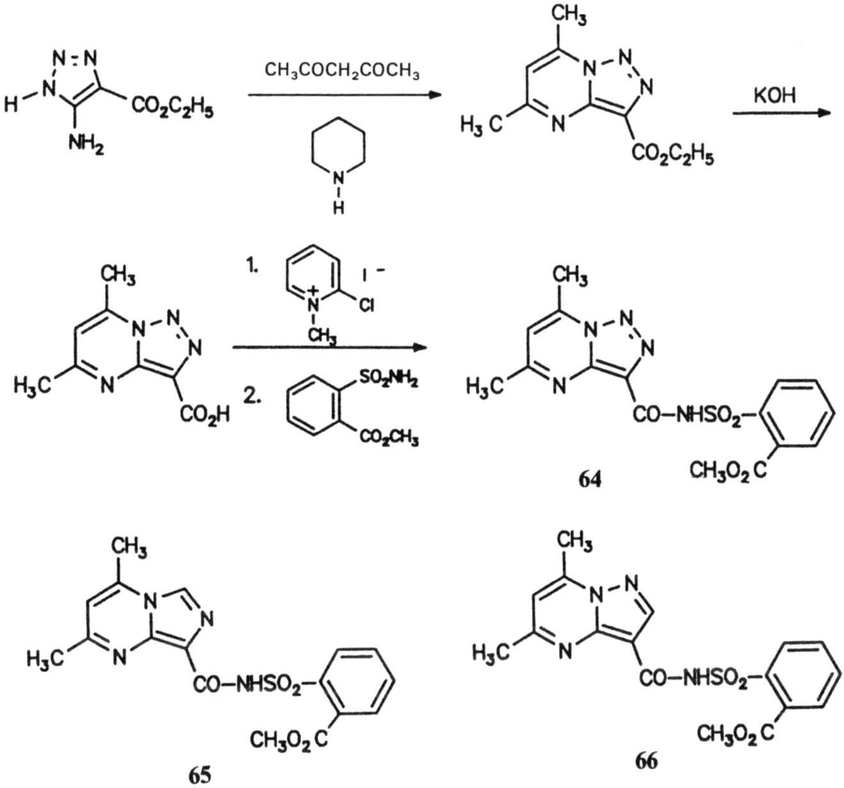

64 65 66

Scheme 29

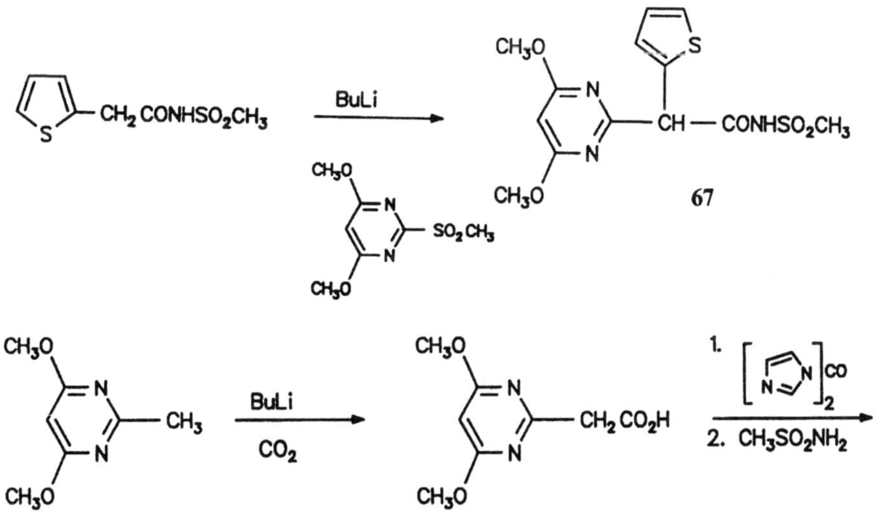

67

Scheme 30

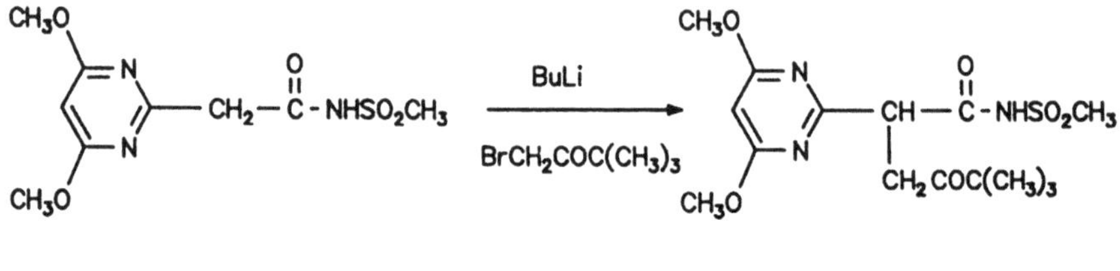

68

Scheme 30 (contd.)

3.2 Hetaryl–N–carbonyl Sulfonamides

Replacement of the aminopyridine/-triazine residue in classical sulfonylureas with substituted triazolinones leads to herbicidal compounds. Triazolinone (**71**) may be built up by condensation of hydrazinoformate (**69**) and lactimether (**70**). Addition of sulfonylisocyanate furnishes the active compound **72** (Scheme 31) [42]. Related patents have appeared containing this type of compounds [43–45].

Analogous pyrazolines also show considerable herbicidal activity. In particular, 3,5,5-trisubstituted compounds are emphasized. The synthesis starts from α,β-unsaturated carbonyl compounds such as **73** which form pyrazolines (**74**) by condensation with hydrazine. It may be concluded by addition of sulfonylisocyanate (**75**) (Scheme 32) [46], or condensation of **78** with sulfonyl carbamate (**77**) (Scheme 33) [47] to give **76** or **79**, respectively. The imidazole derivative (**80**) is synthesized analogously to **79** [48].

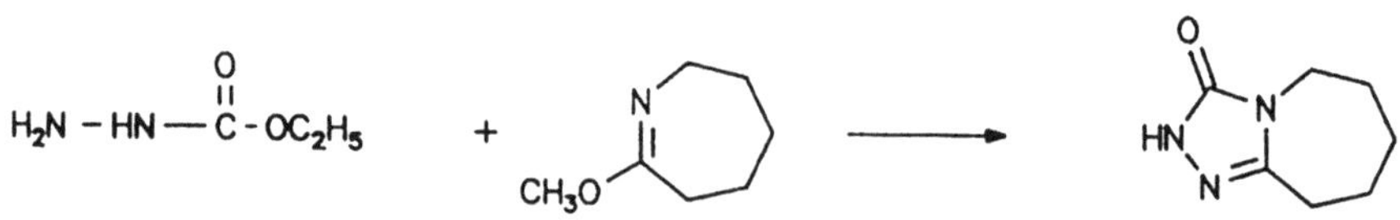

69 70 71

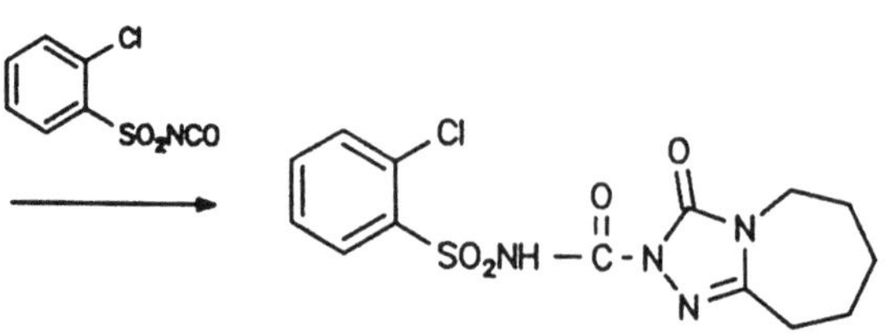

72

Scheme 31

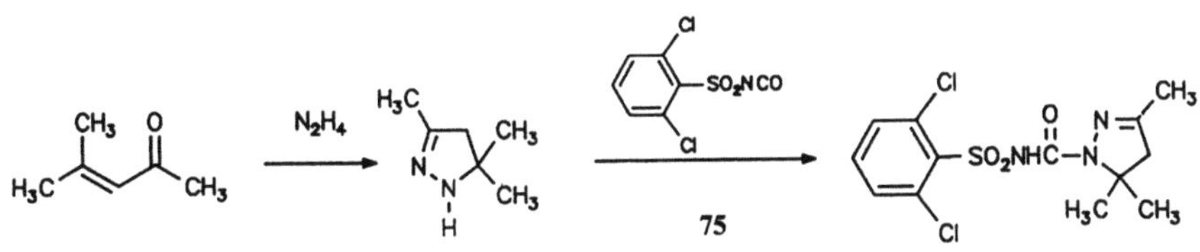

73 74 76

Scheme 32

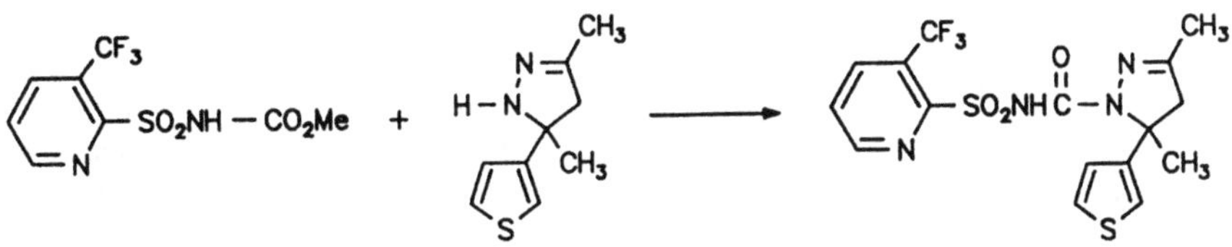

77 78 79

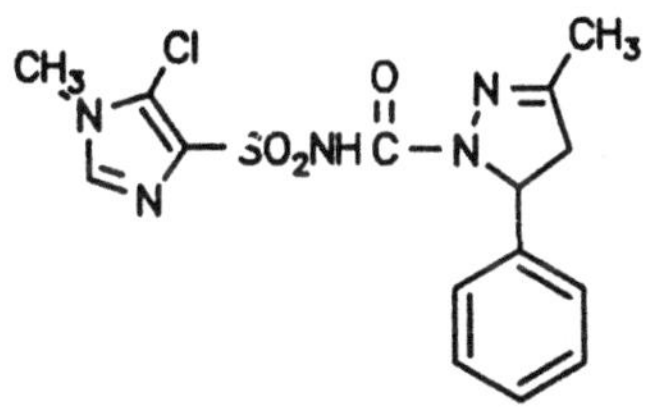

80

Scheme 33

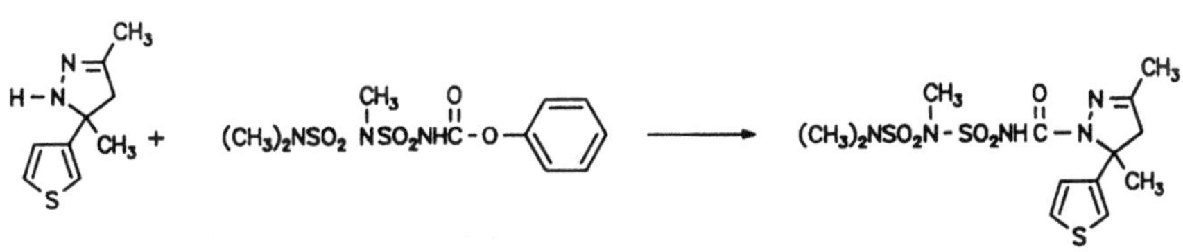

78 81 82

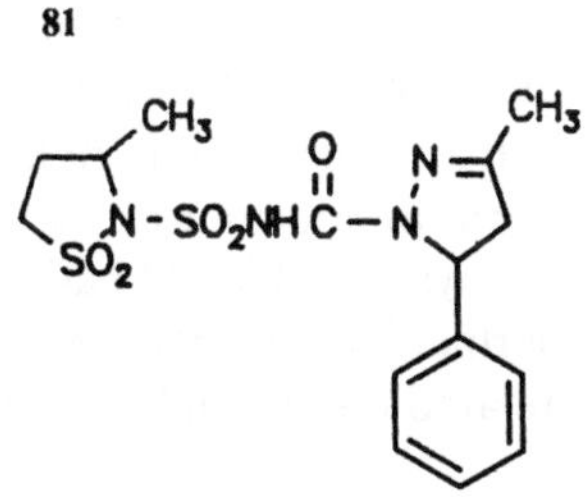

83

Scheme 34

84 **85** **86**

Scheme 35

In addition to arylsulfonamides, sulfamoyl sulfonamides (see section 4) such as **81** have also been coupled to give pyrazolines, e.g. **82** [49] and **83** [50], to produce herbicidal substances as depicted in Scheme 34.

In an early patent [51], simple alkylated sulfonylureas **(86)**, are claimed to be herbicides. They may be obtained by addition of sulfonylisocyanate **(84)** to dialkylamines, e.g. **85** (Scheme 35). Compared with more recent commercial sulfonylureas, however, much higher doses (2.5 kg/ha) have to be applied.

4 Sulfamoylsulfonylureas

In 1983, Willms (Hoechst) reported sulfamoylsulfonylureas (**87**) which are potent ALS inhibitors though they completely lack the aryl moiety of classical sulfonylureas [52].

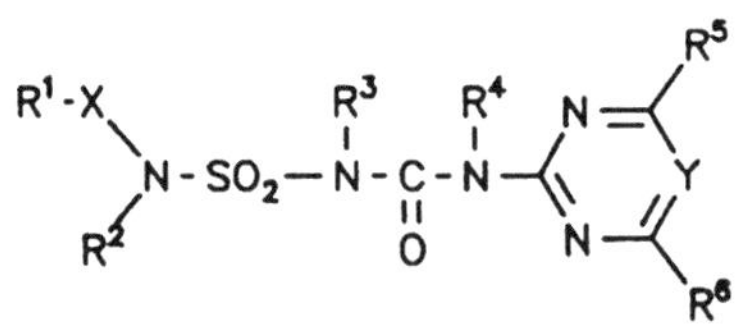

87

Considering the synthesis of the commercial herbicide amidosulfuron (**91**) (Scheme 36), one notes that under the reaction conditions chlorosulfonyl isocyanate reacts primarily as isocyanate. At elevated temperatures, the sulfamoyl chloride **(88)** thus formed rearranges to sulfonylisocyanate **(89)**, which adds to aminopyrimidine (**90**) to form amidosulfuron in excellent yield [53].

Even earlier, simple sulfamoylureas like **92** had been prepared (Scheme 37) [54] which were probably herbicidally clearly inferior to amidosulfuron.

Since then, a variety of related compounds has been claimed which will be reviewed in this section.

To begin with, it was sensible to incorporate the residues R^1, R^2 in **87** into a ring system. Indeed, the sultams **(93)**, which are obtained according to Scheme 38, are herbicidally active [55–57].

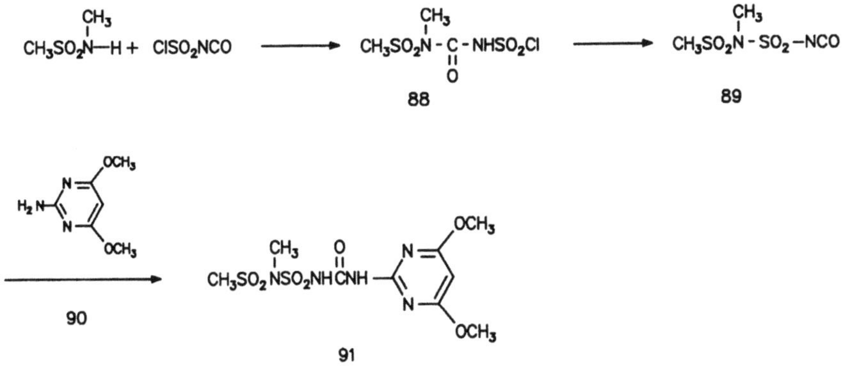

Scheme 36

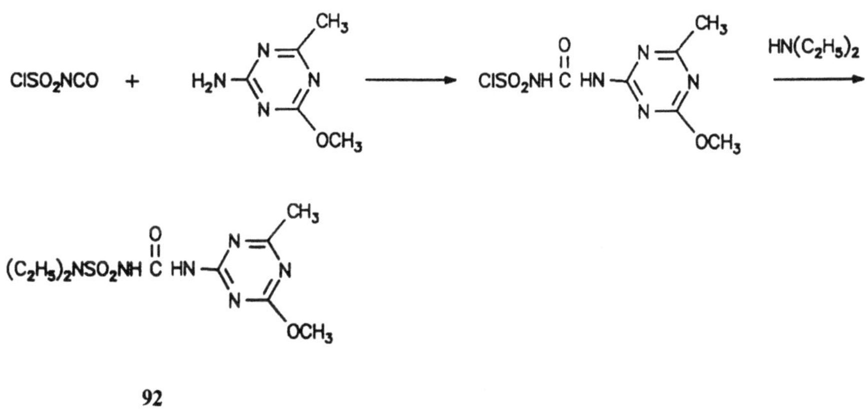

Scheme 37

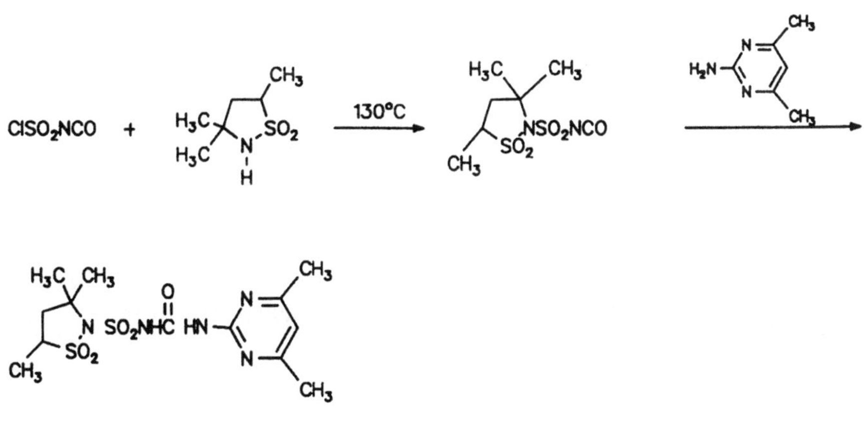

Scheme 38

When a phosphate instead of the sulfamoyl residue is chosen as R′–X the analogous phosphamoylsulfonylurea **(94)** is produced as shown in Scheme 39 [58].

Alkyl-spacered derivatives of amidosulfuron such as **95** (R^1–X = $C_2H_5SO_2CH_2CH_2$) and the corresponding reduced thioethers **(96)** possess herbicidal activity. Scheme 40 displays the synthesis of compounds **95** and **96** [59].

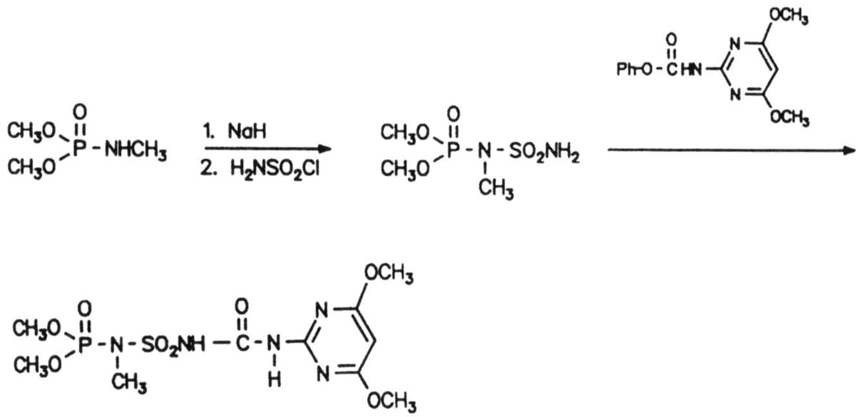

94

Scheme 39

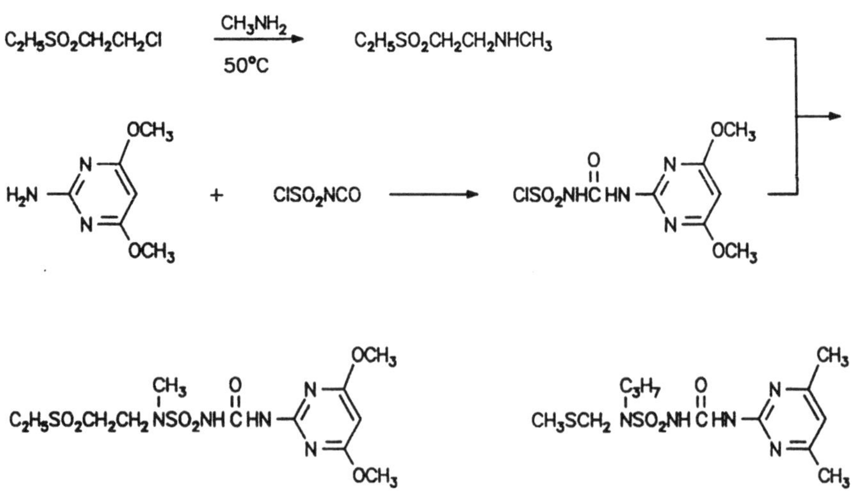

95 **96**

Scheme 40

Sulfonyldiamides (R^1–X = R^2NSO_2), e.g. **97**, **98** [60] and **99** [61] (Scheme 41), can be synthesized according to Scheme 37 or, alternatively, Scheme 38.

Interestingly, carbon analogous compounds such as **100** and **101**, which may be prepared as described in Scheme 42, are also herbicidally active [62, 59].

When the sulfamoyl residue in **100** is replaced with a carboxylic ester function, one obtains herbicidally and growth regulatory substances of type **102** (Scheme 43) [63].

Aside from the variations of R^1–X in **87** which have just been reported, replacement of the alkyl residue R^2 with alkoxy substituents results in promising products. Application of the approach used in Scheme 36 gives rise to the

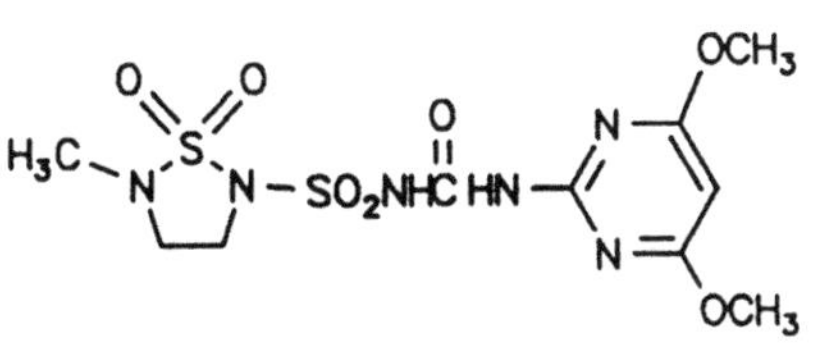

97

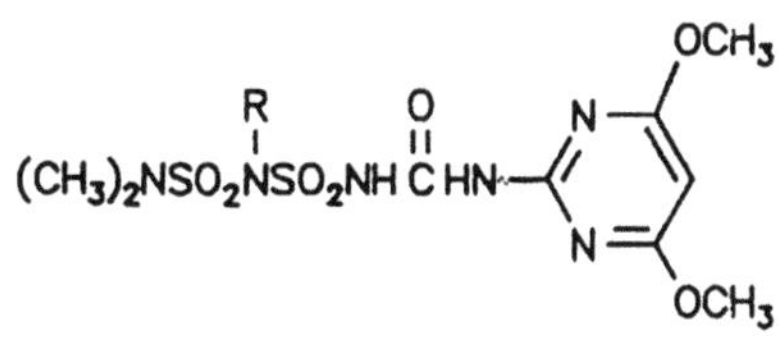

98 R=CH$_3$

99 R= ◁

Scheme 41

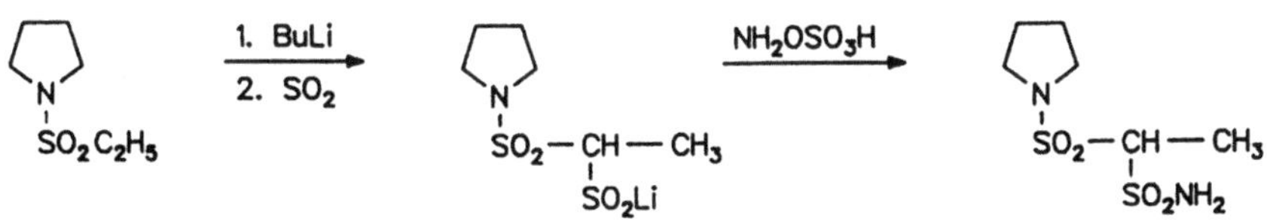

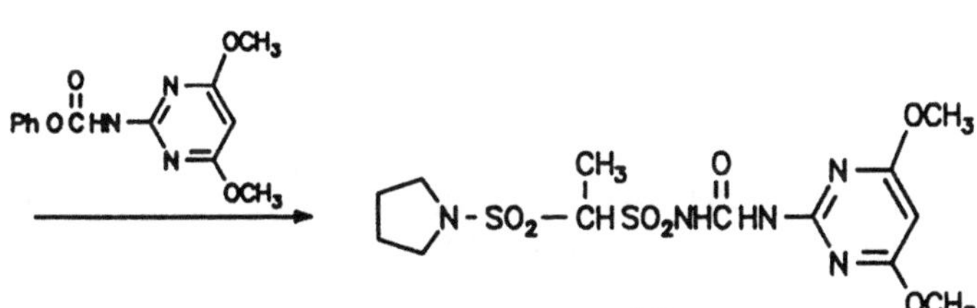

100

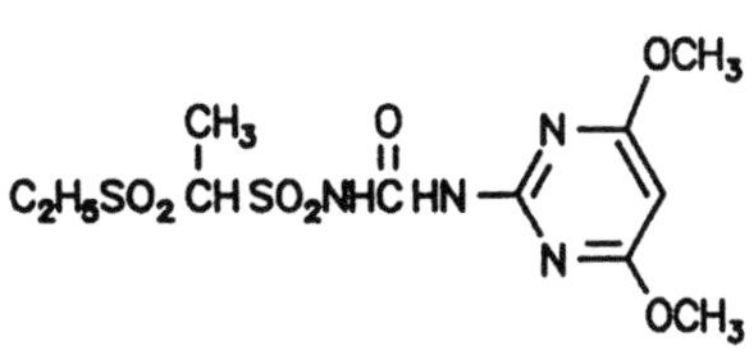

101

Scheme 42

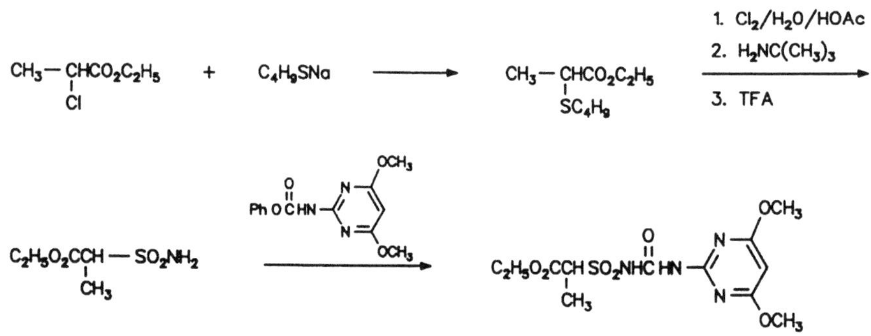

102

Scheme 43

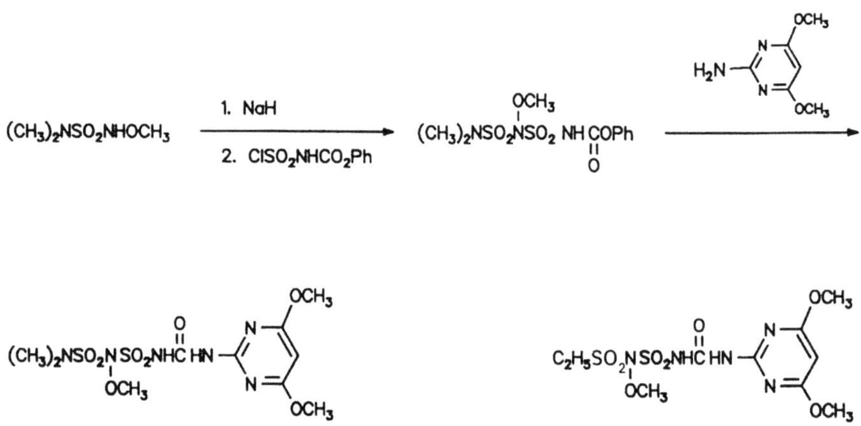

103

104

Scheme 44

N-methoxy derivative **104** [64] or **103**, the sulfamoyl analogue of **104**, which can be prepared as shown in Scheme 44 [65].

Acknowledgement. The authors wish to thank Dr. K.-H. Müller and Dr. A. Plant for valuable comments, suggestions and discussions.

References

1. Takeda: JP 63 091 390 6.10.86/22.4.88
2. Nippon Soda: JP 02 160 782 13.12.88/20.6.90
3. Nippon Soda: JP 02 160 784 15.12.88/20.6.90
4. Bayer: EP 384 244 10.2.90/29.8.90
5. Bayer: EP 330 959 20.2.89/6.9.89
6. Bayer: DOS 4 035 141 6.11.90/7.5.92
7. Schering: DOS 3 737 748 3.11.87/18.5.89
8. Schering: EP 244 948 3.4.87/11.11.87; EP 338 465 15.4.89/25.10.89; EP 246 749 21.4.87/25.11.87; EP 417 875 11.9.90/20.3.91
9. Nippon Soda: JP 02 160 785 15.12.88/20.6.90
10. Bayer: EP 384 250 10.2.90/29.8.90
11. Agro Kanesho: JP 02 304 082 16.5.89/17.12.90
12. Bayer: EP 355 385 15.7.89/28.2.90
13. DuPont: EP 5 986 29.5.79/12.12.79
14. Bayer: EP 413 221 4.8.90/20.2.92
15. Bayer: EP 507 172 23.3.92/7.10.92
16. Bayer: DOS 3 818 040 27.5.88/30.1.89
17. Bayer: EP 121 082 23.2.84/10.10.84
18. Bayer: DOS 3 517 842 17.5.85/13.3.86
19. Bayer: DOS 3 517 821 17.5.85/13.3.86
20. Bayer: DOS 3 431 926 30.8.84/6.3.86
21. Nippon Soda: JP 61 072 783 14.9.84/14.4.86
22. Bayer: DOS 4 017 460 31.5.90/18.4.91
23. Bayer: EP 414 067 10.8.90/27.2.91
24. Bayer: EP 302 378 27.7.87/8.2.89
25. Bayer: DOS 3 602 679 30.1.86/6.8.87
26. DuPont: EP 117 014 3.1.84/29.8.84
27. DuPont: CA 1 242 717 12.8.87/4.10.88
28. Nihon Tokushu Noyaka Seizo K.K.: EP 148 498 27.12.84/17.7.85
29. Bayer: DOS 3 605 549 21.2.86/27.8.87
30. Bayer: DOS 3 431 918 30.8.84/13.3.86
31. Bayer: DOS 3 431 920 30.8.84/13.3.86
32. Bayer: DOS 3 540 840 18.11.85/27.5.87
33. Bayer: US 4 743 294 22.4.87/10.5.88
34. Takeda: JP 63 091 376 6.10.86/22.4.88
35. Takeda: JP 63 091 375 6.10.86/22.4.88
36. Hoechst: DOS 3 826 230 2.8.88/8.2.90
37. Hoechst: WO 91/06 541 19.10.90/16.5.91
38. Bayer: EP 444 286 19.12.90/4.9.91
39. BASF: EP 482 394 18.9.91/29.4.92
40. DuPont: EP 244 166 24.4.87/4.11.87
41. Schering: WO 92/01 677 12.7.91/6.2.92
42. Bayer: EP 341 489 26.4.89/15.11.89
43. Bayer: EP 507 171 23.3.92/7.10.92
44. Bayer: EP 534 266 14.9.92/31.3.93
45. Bayer: DOS 3 936 623 3.11.89/8.5.92
46. Duphar: EP 269 141 7.10.87/1.6.88
47. Nissan: EP 427 174 5.11.90/15.5.91
48. Nissan: JP 03 173 884 8.6.90/29.7.91
49. Nissan: WO 91/13 884 1.3.91/19.9.91
50. Nissan: JP 03 173 887 27.3.89/29.7.91

51. Hoechst: DOS 3 234 450 17.9.82/22.3.84
52. Hoechst: DOS 3 324 802 9.7.83/17.1.85
53. Hoechst: EP 507 093 4.3.92/7.10.92
54. Sandoz: DOS 3 243 533 25.11.82/9.6.83
55. Hoechst: EP 319 689 24.10.88/14.6.89
56. Nissan: JP 03 007 284 10.5.88/14.1.91
57. Nissan: EP 342 456 10.5.88/23.11.89
58. Hoechst: DOS 3 739 691 28.11.87/8.6.89
59. Hoechst: DOS 3 826 609 5.8.88/8.2.90
60. Hoechst: EP 347 788 19.6.89/27.12.89
61. Nissan: JP 03 279 365 30.10.89/10.12.91
62. Hoechst: EP 464 838 5.7.89/8.1.92
63. Hoechst: EP 336 354 4.4.89/11.10.89
64. Hoechst: EP 409 114 14.7.90/23.1.91
65. Nissan: WO 91/06 546 19.10.90/16.5.91

Subject Index

Previously Published Volumes
in Chemistry of Plant Protection

Vol. 1

Sterol Biosynthesis, Inhibitors and Anti-Feeding Compounds

T. Kato
Sterol-Biosynthesis in Fungi, a Target for Broad Spectrum Fungicides

D. W. Krämer
Chemistry of Sterol Biosynthesis Inhibiting Fungicides

K.-H. Kuck, H. Scheinpflug
Biology of Sterol-Biosynthesis Inhibiting Fungicides

D. M. Norris
Anti-Feeding Compounds

Vol. 2

Degradation of Pesticides, Desiccation and Defoliation,
ACh-Receptors as Targets

P. R. Wallnöfer, G. Engelhardt
Microbial Degradation of Pesticides

U. Schoknecht, D. Otto
Enzymes Involved in the Metabolism of Organophosphorus, Carbamate and Pyrethroid Insecticides

R. Richter, D. Otto, H.-J. Mengs
Insecticide Compounds, Acting on the Acetylcholine Receptor of the Insect Nervous System

H. Bermann, D. Martin
Chemical Manipulation of Desiccation and Defoliation and Essential Aspects for the Application and Development of New Chemical Compounds in the Future

Vol. 7

Herbicide Resistance - Bressinosteroids, Gibberellins, Plant Growth Regulators

Vol. 8

Terrestrial Behavior of Pesticides

Vol. 9

Pesticides in Ground and Surface Water

H. Beitz, H. Schmidt, F. Herzel

Occurrence, Toxicological and Ecotoxicological Significance of Pesticides in Groundwater and Surface Water

D. W. Bewick

The Mobility of Pesticides in Soil - Studies to Prevent Groundwater Contamination

C. N. Guyot

Strategies to Minimize the Pollution of Water by Pesticides - A Critical Review to Plant Protection Measures in the Past, Present and Future

M. O. James

Pesticide Metabolism in Aquatic Organisms

G. Mattheß

Fate of Pesticides in Aquatic Environments

M. Häfner

Pesticides in Soil: A German Approach of Predicting Their Movement into Ground and Surface Water

Springer-Verlag and the Environment

We at Springer-Verlag firmly believe that an international science publisher has a special obligation to the environment, and our corporate policies consistently reflect this conviction.

We also expect our business partners – paper mills, printers, packaging manufacturers, etc. – to commit themselves to using environmentally friendly materials and production processes.

The paper in this book is made from low- or no-chlorine pulp and is acid free, in conformance with international standards for paper permanency.

MIX
Papier aus verantwortungsvollen Quellen
Paper from responsible sources
FSC® C105338

If you have any concerns about our products,
you can contact us on
ProductSafety@springernature.com

In case Publisher is established outside the EU,
the EU authorized representative is:
Springer Nature Customer Service Center GmbH
Europaplatz 3, 69115 Heidelberg, Germany

Printed by Libri Plureos GmbH
in Hamburg, Germany